Immigration and Acculturation in Brazil and Argentina

IMMIGRATION AND ACCULTURATION IN BRAZIL AND ARGENTINA

1890–1929

May E. Bletz

First published in 2010 by PALGRAVE MACMILLAN® in the United
States—a division of St. Martin's Press LLC, 175 Fifth Avenue, New York,
NY 10010

Where this book is distributed in the UK, Europe and the rest of the
world, this is by Palgrave Macmillan, a division of Macmillan Publishers
Limited, registered in England, company number 785998, of Houndmills,
Basingstoke, Hampshire RG21 6XS.

Palgrave Macmillan is the global academic imprint of the above companies
and has companies and representatives throughout the world.

Palgrave® and Macmillan® are registered trademarks in the United States,
the United Kingdom, Europe and other countries.

ISBN-13: 978-0-230-10019-0

Library of Congress Cataloging-in-Publication Data

Bletz, May E.
   Immigration and acculturation in Brazil and Argentina : 1890-1929 / May
E. Bletz.
      p. cm.
   ISBN 978-0-230-10019-0 (alk. paper)
   1. Brazil—Emigration and immigration. 2. Argentina—Emigration and
immigration. 3. Acculturation—Brazil. 4. Acculturation—Argentina. 5.
Brazil—Foreign relations. 6. Argentina—Foreign relations. I. Title.

   JV4672.A2B54 2010
   305.9'06912098109041—dc22   2010002672

Design by Scribe Inc.

First edition: October 2010

10 9 8 7 6 5 4 3 2 1

Printed in the United States of America.

CONTENTS

FOREWORD

The idea of this monograph probably started on a trip to São Paulo almost ten years ago. On a rainy afternoon I decided to visit the Museu da Imigração, which used to be the Hospedaria de Imigrantes (Immigrant's Hostel) built between 1886 and 1888, in the Brás district. This hospedaria targeted to accommodate and forward to the plantations the immigrants who arrived here under the responsibility of the provincial government of São Paulo. Nobody had ever heard of the place. I went there once by subway but had to leave, afraid of the neighborhood. The next day, I took a taxi, and, armed with a detailed but as it turned out inaccurate map, about two hours later the friendly driver and I made it. Living in New York City at the time, I was fascinated by the contrast between this neglected Immigrant's Hostel and Ellis Island. That São Paulo had been an immigrant city was very obvious, with its Middle Eastern, Italian, and Japanese restaurants. Apparently, Brazilians treated their ethnicity differently, I decided, and I wondered how and why.

Recently, I Googled the museum and checked their Web site, which did not exist when I was there. A professional site, with a part in English, praises the museum's importance. The site claims the museum retains ship boarding lists, recording logs of sponsored immigrants, "calling" letters (to bring parents and relatives to Brazil), legal proceedings of colony units, as well as personal documents and diaries donated by some immigrants plus eight thousand photographic images, among negatives and originals. Furthermore, the site states, several immigrants and their offspring, organized in clubs and communities, have elected the building of the memorial as a reference and meeting place, strengthening the ties between the past events and the present-day use. The Sector of Oral History of the Memorial do Imigrante is a collective testimony by immigrants from more than sixty nationalities, who arrived here in search of a better life.

The point is not so much whether the memorial in São Paulo will succeed in becoming a tourist attraction similar to Ellis Island, but rather, how cultural practices of identity formation continually shift. After years of neglect, the memorial has been rediscovered by Paulistas who want to officialize their immigrant origin. But nowhere was it mentioned that though that particular neighborhood, Bráz, started out largely Italian, today it is mostly populated, according to the taxi driver, by newer immigrants, *nordestinos*, people from Northeast Brazil. This continual recreation of traditions is the point of departure for this work.

Libraries have been written on immigration from a sociological and historical perspective, rarely from a cultural studies perspective. I discovered I had to come up with different keywords and knit together a corpus. Among the most helpful books I used were the works of Beatriz Sarlo on Argentine modernity and Hugo Vezetti's work on Argentine psychiatry, in which he explains the relation between immigration; the outsider; and the fear of contamination. Both works include discussions on people such as Ricardo Rojas and Ramos Mejía.

On the Brazilian side, I was asked to review Jeffrey Lesser's *Negotiating National Identities*, which helped me tremendously in envisioning this project. Lesser taught me to see how cultural practices, including racist discourses and actions, are neither reducible to nor intelligible in terms of class exploitation alone. In short, these "whitening" practices had a reality of their own and cannot simply be reduced to an economic base.

Obviously, there are gaps in this study that need to be filled by future scholars. A wise outside reader for Palgrave recommended I abstain from discussing Sarmiento's works in great detail. Sadly, I had to cut out the part on the wonderful Roberto Arlt, whose enigmatic prose deserves much more space than I would have been able to give him in this work. I chose certain representative intellectuals, based partly on availability, relevance, and my own taste, but I am fully aware that many others could be mentioned as well.

I have many people to thank for their help with this project. Jane Koustas and The Humanities Research Fund at Brock University enabled me to do research in both Rio de Janeiro and São Paulo in the summer of 2006. Originally, this work started as a PhD dissertation I wrote at the Department of Spanish and Portuguese at New York University. I am deeply grateful for the valuable comments and suggestions by George Yudice of the University of Miami, Mary Louise Pratt, Gerard Aching, Marta Peixoto, and Ana Dopico of New York University. A very big thank you to Sônia Roncador of the University of Texas at Austin for all her enthusiasm and for introducing me to the works of Júlia Lopes de Almeida; to Jason Borge of Vanderbilt University for showing me how to present my project to publishers; and to Joseph Pierce, who did a wonderful job editing the manuscript. I also thank my editors at Palgrave Macmillan—Julia Cohen and Samantha Hasey—for their enthusiasm and professionalism. And to all my friends and family who have supported me over the years—you know who you are—thank you.

INTRODUCTION

OVER THE PAST TWO DECADES, LATIN AMERICAN LITERATURE and culture have been revised by critics attempting to reflect the continent's plural, heterogeneous, and multifaceted society. Yet, to this day, little attention has been paid to the way in which immigrants fundamental to this revision have participated in the formation of modern Latin American identities. In fact, most studies have provided either a strictly economic analysis of the effects of immigration or some essentialist notion of foreignness. This being the case, my purpose with this book is to shed light on the reciprocal nature of the processes of representation and identity formation that took place between national and immigrant groups in Argentina and Brazil from 1880 to 1929 and to explore how migration was constructed as a metaphor for modernity in both countries.

I start from the premise that after the independence and unification of its federal provinces in the second half of the nineteenth century, Argentina was seen as capable of becoming a "modern" country by encouraging foreign immigration.[1] Perhaps the most influential immigration policies were formulated by Bernardino Rivadavia, who in 1818 stated the following:

> El aumento de población no es sólo a ese Estado su primera y más urgente necesidad, después de la libertad, sino el medio más eficaz, y acaso único, de destruir los degradantes hábitos españoles y la fatal graduación de castas, y de crear una población homogénea, industriosa y moral, única base sólida de la Igualdad, de la Libertad, y consiguientemente de la Prosperidad de una nación.[2]

[For this State, population increase is not only its first and foremost necessity after freedom, but also the most effective, and perhaps the only way to destruct the degrading Spanish customs and
the fatal graduation of castes, and to create a homogeneous, industrious and moral population, the one solid foundation of Equality,
Liberty and consequently of the Prosperity of a Nation.]

Insofar as the legal definition and political conceptualization of
the citizen enfranchise the subject who inhabits the national public sphere, the concept of the abstract citizen—in which each is
equal—is defined by material labor conditions and the inequalities
of the property owned. With regard to immigration policies in the
United States, Matthew Frye Jacobson notes how scholarship has
generally conflated race and color, without keeping in mind that,
in the nineteenth and early twentieth centuries, one could easily be
"white" and racially distinct from other "whites."[3] It is this history
of "becoming white" combined with "becoming a national subject"
that I trace in this book.

We should not blindly assume, however, that the immigration
policies of the United States can stand as a "model" for Argentina
and Brazil to simply imitate. If read carefully, the texts on foreign
immigration and race relations that were written and discussed at
the beginning of the last century paint a completely different picture. By the end of the nineteenth century, as we shall see, the
meaning of "nation" had gradually moved from a voluntary association to an ethnic concept of a new national race and character.
However, the idea of a "national race" was riddled with problems
for Latin American intellectuals, not only because of the high
level of racial mixing when compared to the United States, but
also because their original "whiteness" was perceived to be different
from that of their northern neighbors. As Matthew Frye Jacobson
argues, "[To] miss the fluidity of race itself . . . is to rely on a mono-
lithic whiteness, and, further, to cordon that whiteness off from
other racial groupings along lines that are silently presumed to be
more genuine."[4]

Even the "white" origins of Latin America were questionable,
and founding fathers like Juan Batista Alberdi and Domingo Faus-
tino Sarmiento were quite persistent in their admiration of the
"Anglo-Saxon" rather than the "Latin" races. Immigration was

seen as a crucial agent of modernization by the Argentine elite and was linked to notions of cultural as well as economic and political progress. Meanwhile, in Brazil, the Portuguese and their colonial traditions were increasingly scorned, and in both countries medical discourses flourished toward the end of the nineteenth century, offering a way to organize and institutionalize a national identity. Thus, the differences between desirable and undesirable immigrants were assumed to be found in the body, in what were then taken as racial characteristics.

One of the main ways in which Argentine intellectuals of the mid-nineteenth century expressed their concern for their country's racial makeup was their call for specifically Northern European immigrants. Behind this project was the desire to populate and make productive Argentina's vast pampas, the rangelands, and agricultural zones of the interior, in the hope that the remaining indigenous peoples would be absorbed into the national body and that the mestizos and criollos would be counterbalanced with "superior" Northern European stock. Brazilian authorities, in turn, felt the need to catch up with their "whiter" neighbors and promoted a similar type of immigration.

Which institutions and discourses were created to "produce" these new citizens? In other words, how did the Brazilians and Argentines that had arrived before the period of mass immigration employ the figure of the immigrant in constructing a national culture? What criteria were these immigrants expected to meet, and to what extent were they allowed to do so?

The texts I examine in this work are all drawn from the period between 1880 and 1920, a period in which conflicting attitudes over immigration and immigrants were particularly intense and regularly interspersed with the rhetoric of modernity and progress and the desire for the birth of a new age. European immigration came to an end around 1930 and was resumed at a more modest level after World War II. Eventually, immigrant cultures were "absorbed," creating new ideological frameworks that enabled immigrants to affiliate themselves with their new home country and national elites to consider them something more than "strangers."

These texts are all profoundly urban in character. This is no coincidence, since, as Caren Kaplan points out, modern cities, and especially major ports, function as crucibles where identities are formed, transformed, and fixed.[5] Deeply connected to the story of immigration is the history of the modernization of Buenos Aires, which in its turn was imitated to a certain extent by Rio de Janeiro and São Paulo. These new "immigrant identities" were certainly not always self-chosen, welcome, or advantageous to the newly arrived, but they did play a role in the deployment of political interests on the part of state institutions and subsequently, on literary and artistic canons as well.

This book deals with sociological and historical essays, treatises on education, as well as fictional writings, which range from so-called positivist novels to avant-garde texts. All try to come to terms with the ambiguous figure of the immigrant in rapidly urbanizing societies. In my opinion, such a corpus is made possible when certain connections between immigration, citizenship, and national belonging undergo a transformation that requires new definitions of national citizenship.

The relative neglect of immigration and ethnicity in Latin American Cultural studies is, as I hope to show, more a lack of interest in this type of scholarship than a lack of material. While local elites, schooled in the positivist and eugenic philosophies of the late nineteenth century, disdained any rural or religious traditions, they would also not accept the diffusion of nationalisms, be they Italian, Spanish, German, or Yiddish. Racially speaking, these immigrants were perhaps "civilized" people, but the original policies as formulated by Argentina's founding fathers implied that they had to fulfill a civilizing "national mission" as well. European immigrants discovered all too soon they could benefit from local racial policies and often only acquired an "ethnic consciousness" and a "white" skin color after coming to their host country.

IMMIGRATION AND ITS "OTHERS"

Roberto Schwarz, in his chapter "Nacional por subtração," describes Brazilian thought on national identity as employing a similar process of elimination.[6] As a former colony and "underdeveloped" country, Brazil found that each generation of Brazilian intellectuals was typically more interested in the recent theoretical productions of the European metropolis than that produced by previous generations in their own country. As Brazilian intellectuals became increasingly conscious of the "inadequacy" of imported models, many went to the opposite extreme of believing that in order to attain a substantive intellectual life, it was sufficient simply not to reproduce metropolitan tendencies. The Brazilian intellectual, according to Schwarz, now saw the recovery of a "genuine" national culture as a reconquest, an expulsion of foreign invaders, and an elimination or excision of all that was not autochthonous. The residue of this subtracting operation would be the authentic substance of the country. Any attempt to describe Brazilian cultures, Schwarz argues, should try to include all of its variations, including those of the poor and marginalized, instead of excluding them in order to find or create this supposedly genuine Brazil.

Jeffrey Lesser takes up the challenge pitched by Schwarz and tries to modify the widely held thesis that elite conceptions of national identity were predicated on the elimination of ethnic distinctions.[7] Rather than taking the process of elimination for granted, Lesser argues, scholarship should focus on the ways in which notions of national identity were continually contested and negotiated.

In Argentina, immigration debates first revolved mainly around which of the European "races" and which social classes best represented Argentine nationality. In his analysis on Argentine immigration in *Cousins and Strangers*, José Moya criticizes contemporary scholars for focusing too much on the "heroic" and the supposedly pure, essential, non-European "otherness" in Latin America.[8] While I certainly hope that concern about human-rights abuses will stimulate scholarship on, for example, indigenous groups of the Amazon forest that goes beyond their supposedly "exotic" character, I do agree with Moya that within Latin American scholarship, we too often see a rhetoric seemingly in quest of a pure "home country," which is still too often deployed as an ahistorical, metaphorical, and often sentimental space. I situate this work within the theoretical framework of diaspora studies by Edward Said, Matthew Jacobson, and Caren Kaplan, among others. These studies do not focus on Latin America, but their use in this study, however, will always be placed within a rigorous historical and geographical framework.

THE LANGUAGE OF EUGENICS

As Nancy Stepan shows, any discussion of Latin American racial policy in the nineteenth century is intrinsically linked to the era's debates on eugenics, theories that specific traits were acquired via local human and climatic environments.[9] Since "racial purity," from both a biological and a cultural standpoint, was (and is) necessarily an illusion, a product of nostalgia for an imaginary era of harmony and homogeneity, the mixing of conceptual categories like "nation" and "ethnicity" was often described with racial language. The shifting attitudes regarding nationality and ethnicity were frequently revealed in discussions about the desirability of certain immigrant groups. Especially damaging to the self-image of Latin America was the scientific view of racial hybridization, universally condemned by European biologists as the cause of Latin American degeneration. Latin American intellectuals were only too prone to project onto themselves these negative judgments of the outside world. Yet, as Stepan illustrates, there were also limits to the degree to which Latin Americans could apply to themselves a thoroughly racialized view of the capacity of their respective countries for civilization and progress.[10] They asked whether racial mixture was always a sign of inferiority or a cause of national decay; whether hybridization could have more positive biological and social meanings; and whether it should be encouraged as a biological process of nation formation, allowing the emergence of a national homogeneous population through a process of racial fusion. These efforts to reevaluate the national self were carried out in the name of race, not in rejection of race as an explanatory variable of history. How was Brazil to fill its vast empty spaces in the inland or Argentina its pampas? How could these countries exploit their abundant natural resources without healthy, working populations?

ETHNICITY AND CITIZENSHIP

The postcolonial understanding of national and cultural identities is predicated on the belief that they are the result of a historically negotiated process and can, therefore, only be understood in relation to each other. The awarding of citizenship to large numbers of immigrants signals a new concept of nationality. In this sense, I am particularly interested in Michel Foucault's studies on the relation between State and racism.[11] In these studies, Foucault explains that the modern state requires a continual production of an "internal enemy" in order to justify its disciplinary functions. The modern state is in charge of the "health" of the population and will need the (re)creation of internal differences that will permit the control of individuals and groups. These differences signal and define certain people and groups as undesirables, unfits, and even enemies whose elimination will restore the threatened national health. The production of an internal frontier as an instrument to distinguish between individuals, to determine and separate "identities," and to create divisions within society will be, according to Foucault, one of the principal functions of the modern state. As governmental practices have addressed themselves in an increasingly immediate way to "life," in the form of the individual detail of individual sexual conducts, individuals have begun to formulate the needs and imperatives of that same life as the basis for political counterdemands. Biopolitics, explain Graham Burchell, Collon Gordon, and Peter Miller in their study on Foucault, thus provides an example of the ways in which the terms of governmental practice can be turned around into foci of resistance, or the ways the history of government as the "conduct of conduct" is interwoven with the history of dissenting "counter-conducts."[12]

Using Georg Simmel's concept of the "stranger" as the person who comes today and stays tomorrow, a potential wanderer,[13] I argue that the figure of this stranger becomes a terrain permanently reinscribed by distinct and competing narratives of national identity from which different, potentially dangerous figures emerge. These figures embody what needs to be classified, separated, and perhaps even eradicated in order for the nation to survive.

I follow the suggestion of Bonnie Honig that the figure of the immigrant must be seen in a double sense: as both a source of "enrichment" and an element that can "corrupt" the adopting country.[14] Immigrants are inherently perceived in an ambiguous way; on the one hand, "they" bring to "us" diversity, energy, and a renewed appreciation of our culture. Yet, there is also a fear of what they will do to us: dilute our common heritage. Here the immigrant's choice to come to our land endangers our sense of who we are. Women and men who move between cultures, languages, and the various configurations of power and meaning in complex colonial situations develop unique responses to questions of identity and writing.

In an elaboration of Edward Said's notion of affiliation,[15] Caren Kaplan correctly notes that when questions of identity are raised within a framework that deals with displacement, notions of immigration and diaspora are seen as either historically or culturally distinct.[16] Historicizing displacement, she argues, leads the critic away from nostalgic dreams of "going home" to a mythic, metaphysical location and into the realm of theorizing a different way of being at home. In such a scenario, immigrants are perceived as, treacherously, replacing one nationalist identification for another, while émigrés confound territorial and essentialist nationalisms in favor of transnational subjectivities and communities. The modes of displacement that come to be attached to the figure of the immigrant, therefore, run counter to those most valued by Euro-American modernism. Rather than embodying the desire to return to a lost origin, the immigrants are represented as eager to reject that origin and are suspected of putting material gain above their place of birth. Without any doubt, Said is deeply committed to displaced individuals, yet he still handles a strict dichotomy between filiation—a "natural" link to one's family and country—and affiliation, a nonfilial tie.[17] He fails to examine the idea that there is no "natural" link between a place and a people. Ultimately, Kaplan points out, both filiations and affiliations, to use Said's terminology, are learned, created, recalled, or forgotten in everyday history. A necessary alteration to propositions like Said's would require us to see filiations as those bonds that are naturalized through the discourses that differentiate them from those bonds that are imagined as artificial, that is, as affiliations.

Another researcher committed to examining notions of displacement is James Clifford, who analyzes the acting out of relationships rather than the departing from pregiven, "natural" forms.[18] He examines identity and community formations that negotiate the historically produced tension between movement and dwelling, focusing on what the diasporic dimensions are that create links between people in diverse locations, constituting identities that do not reproduce nationalisms. Transplanted, the individual is transformed, and the "I" is no longer a speaking subject with a clear history and a distinct voice, but rather becomes the composite product of historical antinomies and contradictory impulses.

The notion of a "composite I" when talking about immigration is particularly useful. It is all the more surprising, then, that Clifford, writing in Southern California, never mentions immigration as part of this process. By ignoring class differences and failing to examine his own surroundings, Clifford falls into the modernist trap of assuming an opposition between exile, which we associate with something noble and tragic, and immigration, which seems plebeian and even disloyal in comparison. By exclusively referring to a "bad" host country where the displaced subject is "misunderstood" without acknowledging the possibility of a desire to assimilate, Clifford thus continues to privilege one's birth location as the place where one "truly" belongs.

For these reasons, I intend to discuss historically negotiated national subjects—which, of course, we all are—rather than to depart from some exclusivist notion of diaspora. I therefore adhere to the Foucauldian idea that knowledge, discourse, and power are strongly associated. Citizenship requires that the subject deny his or her particular private interests to become an "abstract citizen" of the political state. Citizenship, the history of immigrants tells us, is never an operation confined to the negation of individual "private" particulars. On the contrary, it requires the negation of a history of social relations that publicly racialized groups are obliged to "forget." Furthermore, it requires acceding to a political fiction of equal rights that is generated by the denial of certain alternative histories in order to create the ontology of the nation. It is from this consideration that Foucault develops his concept of biopower, which is related to concrete constraints on major aspects of the human body such as movement and especially sexuality. Immigrant communities can thus be seen as crucial sites where the terms of membership in the national body are contested, policed, and ultimately defined. What exactly constitutes a "foreigner" is a matter not simply of law but of ideological processes and debates. It is precisely these debates that I am interested in tracing.

Lisa Lowe, in *Immigrant Acts*, points out that citizens inhabit the political space of the nation, a space that is, at once, juridically legislated, territorially situated, and culturally embodied.[19] Although the law is perhaps the discourse that most literally governs citizenship, national culture—the collectively forged images, histories, and narratives that place, displace, and replace individuals—powerfully shapes which individuals will be accepted as citizens, where they live, what they remember, and what they should, though perhaps cannot, forget. Therefore, I chose an interdisciplinary approach to study mass immigration in Argentina and Brazil. Economics, law, and racial and political theories cannot be privileged over the realm of "culture." Culture not only is the medium of the present—the imagined equivalencies and identifications through which the individual invents lived relationships with the national collective—but also is simultaneously the site that mediates the past, through which history is grasped as difference, fragments, and lashes of disjunction.

How was this process of acculturation and negotiation eventually resolved? My hypothesis is that a concept of "national identity" in both Argentina and Brazil only established itself after the periods of mass immigration, that is to say, in the 1930s. Eventually, the notion of a "foreign settler" faded away and was replaced by something else. On the one hand, Argentina, with its largely white local population and equally white immigrants, relegates ethnicity to class distinctions and creates a pliable "*gaucho*" figure, with whom all Argentines, old and new, rural, but especially urban, are given the chance to identify, thus promoting their newly invented "rural origins."

In Brazil, on the other hand, due to its large population of African descent, immigrants were categorized under a system of racial labels, and a cosmopolitan *mestiçagem* was eventually installed as official discourse, replacing earlier deterministic ideas of race and providing the institution that would later allow for a Brazilian "national type." It is particularly interesting to note that in the southern region and the São Paulo area, where the immigration process was relatively similar to that of Argentina, the "*bandeirante*," an equally mythical counterpart to the Argentine gaucho, eventually emerged.

In Chapter 1, "In Sickness and in Health," I focus on the specific challenges that eugenic thinking presented Brazilian elites. Doubts about the country's racial identity had long been reinforced by racist interpretations of Brazil from abroad, and for positivist French thinkers, Brazil was held up as a prime example of the "degeneration" that occurred in a racially mixed, tropical nation. Early European thought saw the tropics as a healthy place, but this idea changed over time and eventually the real and very high mortality rate of Europeans in Brazil, who lacked the immunity to certain diseases that native people had acquired, caused many to view the country with dread. This lead to the notion that some races were naturally suited to the tropics, while others, particularly Europeans, had better stay away.[20]

Positivist ideas in Brazil coincided with the abolition of slavery in the country in 1888, legislation that encouraged the migration of many former slaves from rural areas toward the cities. Around the same time, foreign immigration began in earnest, and both movements combined caused a wide-ranging process of destabilization of traditional Brazilian society and culture.

The specific encouragement of European immigration by the end of the nineteenth century had become national policy in Brazil. Representative of the government's optimistic attitude is João Cardoso Menezes e Souza's *Theses sobre colonização do Brazil*, published in 1875.[21] Soon however, dissent was heard from such intellectuals as Raymundo Nina Rodrigues, who proposed a strict racial segregation in the country.[22] An early preoccupation with race would lead Nina Rodrigues to the determinism of nineteenth-century anthropology and racial "science." Throughout his writing, he questions the relationship between race and pathology, and he undermines the idea of human agency as a factor in controlling the progress of the population.

In contrast, Eduardo Prado defended the Brazilian monarchy against the First Republic, which, according to him, was based on a most unfortunate imitation of liberal ideas imported from the United States.

I conclude the chapter with readings of Aluísio de Azevedo's *O cortiço* (1890) and Júlia Lopes de Almeida's *A casa verde* (1932, written in 1899), tying the novel's plots to racial debates of that period. By reading *O cortiço* with the sociological texts of Nina Rodrigues, Prado, and Souza, I suggest the novel can also be read as a reaction against the positivist ideologies of *branquemento* or racial "whitening" that stimulated European immigration. Almeida's novel is by no means indicative of the literary quality of her other works. *A casa verde* is a mystery novel, with a highly implausible plot. However, it is probably her work that most clearly shows the unresolved tension between her rejection of deterministic thought and her strong belief in European, particularly British, superiority.

In Chapter 2, "Purifying the Urban Landscape: Processes of Immigration, Acculturation, and Resistance in Buenos Aires," I begin with a discussion of a sociological text by José Ramos Mejía, *Las multitudes argentinas* (1899), and two novels, one by Julián Martel, *La bolsa* (1891), and the other, *Stella* (1905), by Emma de la Barra, which offers readers an unusually positive portrayal of the female immigrant, a figure who was generally seen as either an "antimodern" figure or a treacherous prostitute. All of those works address concerns raised by Sarmiento about immigration. Francine Masiello observes that concomitant with the emergence of a paranoid, nationalist subject, anthropological discourse gains legitimacy by providing the institutions of power with a therapeutic and prophylactic perspective on the "dangerous" classes. In that sense, she explains, the foreign virus operates as a self-constituting other, as the condition that, paradoxically, enables the constitution of the healthy national subject.[23] The therapeutic gaze of the anthropologist responded to such a crisis by representing and ordering the flow of immigrants to Argentina. In many ways, the tableaux of criminal traits were the "scientific" effect of a sinister hermeneutics that interpreted the physiological traits of ethnic difference as signs of social and moral disorders. It is precisely in the transition period between the liberal state and another more "modern" society that the concept of a variety of "inferior races" starts being used more frequently. In this, we can see how Argentine sociology, criminology, and anthropology were modeled after positivist social and natural sciences. The emergence of a "Hispanic" conservative nationalism must be seen, I suggest, as a response to working-class cultures in factories, urban slums, and tenement buildings, while the desire to "purify the city" should be seen as a response to the chaotic urban developments in Buenos Aires.

In Chapter 3, "Negotiating New Identities: Argentina of the Centennial," I examine how the positions of immigrants and gauchos shifted after the days of Sarmiento's and Alberdi's pushes for European immigration. Rather than the immigrants being been seen as the element of civilization, which would tame "barbarian" gaucho culture, the gaucho was now invoked as the true Argentine who would keep the barbarous immigrants at bay. I argue that the emergence of an ethnocultural understanding of "Argentine nationhood" coincided with, and indeed was in large part precipitated by, a massive influx of European immigrants. While deploring the newcomers as a threat to the collective race or character, cultural nationalists and their sympathizers accepted that immigration was inevitable and believed that the incoming masses should be assimilated or "Argentinized" as completely as possible.

Since the majority of immigrants came from rural areas, it is not surprising that they should initially dream about having their own land upon arriving to their new country. This dream coincided with the urban elites' fears of unionization in industrial areas of Buenos Aires. Therefore, gradually an idea of a "pastoral utopia" was being developed, an idea that can be seen in an immigrant's conduct manual like the *Manuale dello emigrante italiano all'Argentina*, originally written by Arrigo De Zettiry.[24]

The rapid changes caused by immigrants and their descendents were, of course, most directly perceived in Buenos Aires. Concerned elites therefore started reevaluating their previously dismissive opinion of the countryside, which maybe represented the true Argentine spirit. The idea of a lost agrarian, antimodern utopia is prominent in the nationalist thinker like Ricardo Rojas.

Alberdi and Sarmiento had thought that native Argentines could acquire, through simple contact or more formal education, desired Anglo-Saxon traits. As mentioned previously, the meaning of "nation" had gradually moved from a voluntary association to an ethnic concept of a new national race and character. Cultural nationalists like Rojas resolved this dilemma by focusing on supposedly unique cultural and historical elements rather than biological traits. Education thus becomes a key issue in the acculturation process; history and "national character" were to be taught in state schools. José María Ramos Mejía was confident that the immigrants' offspring would gradually improve, and be more intelligent and attractive, due to the better food, climate, and education. Alternatively, for Ricardo Rojas, history was a laboratory for a future national moral standard, a patriotic lesson that welcomes and educates carefully selected citizens. Probably the best-known text written by an immigrant in Argentina is Alberto Gerchunoff's *Los gauchos judíos* (1910). Gerchunoff cleverly manipulates nationalist, xenophobic thought by suggesting that it is in the countryside, far away from the city with its labor unions and anarchist ideas, where the immigrant can truly learn to love the Argentine soil and truly become Argentine.

In my fourth and final chapter, "Brazil and Its Discontents: Romero and Torres," I examine the creation of a "national Brazilian type" with whom Brazilians and newcomers alike are asked to identify, albeit in a utopian future. My discussion begins with Sílvio Romero's 1906 attack on "the German danger," *O allemanismo no sul do Brasil, seus perigos e méios de os conjurar*, followed by a brief discussion of Alberto Torres's main arguments in which he denies the importance of a "scientific race," or rather substitutes for that category a "national character." Graça Aranha's novel *Canaã*, published in 1902, plays with these fears of German separatism, racial degeneration, and national character, and the author offers a possible solution by referring to a chronologically and geographically mythical place, a promised land, Canaan. *Canaã* is one of the first works that engaged in this search for a perfect race through *mestiçagem*. It is this idea, rather than some supposed stylistic or artistic program, I argue, that the younger *modernistas* learned from Graça Aranha.

I will then conclude this final chapter by examining the "cosmopolitan" way in which immigrants were imagined in Brazilian "urban writing" of the 1920s, focusing on *modernismo* in São Paulo. I suggest that although the *modernistas* were certainly deeply involved in new, shocking ways of expressing their reality, they also form part of a larger debate concerning which "national" and "foreign" cultures are "suitable" for import or export. I will analyze the figure of the immigrant in the *modernista* writer Antônio

Alcântara Machado in his collection of short stories *Brás, Bexiga e Barra Funda* (1927). These narratives portray the varying degrees and methods of integration into *Paulista* society achieved by Italian immigrants and their offspring, especially focusing on their children in this society in transformation. More specific to the national context in which this work was created, Alcântara Machado also creates his own theory of race.

CHAPTER 1

IN SICKNESS AND IN HEALTH

THE NATION'S DISEASE

JEFFREY LESSER[1] REMARKS THAT NOTHING ALLOWED POLITICIANS TO see their country as a "racial laboratory" more than the presence of immigrants. Immigration played an important role in public policy from at least 1850, when it became clear that slavery would not exist long into the future. Although most elites did not seek to use immigrants as a replacement for the largely exterminated native population (as was the case in Argentina), they did assume there was a high correlation between the influx of immigrants and social change. Jeffrey Lesser mentions that between 1872 and 1949, around 4.55 million immigrants entered Brazil and these immigrants challenged simplistic notions of race by adding a new element—ethnicity—to the mix.[2]

Some intellectuals and politicians sought "pure" European immigrants who would recreate the Old World in the New. These "whitening" immigration policies provoked sharp reactions among Brazil's elites from across the political spectrum, while often at the same time subscribing to deterministic notions of racial superiority. In this chapter, I will outline some of the main ideas surrounding immigration in the late nineteenth century, from the strongly deterministic ideas of Menezes e Souza, to Nina Rodrigues, to the monarchist Eduardo Prado, to the abolitionist Aluísio de Azevedo.

TROPICAL DEGENERATION

According to Nancy Stepan, Brazilians saw themselves as a racially diverse though a largely dark-skinned people, the product of generations of intermixing between Indigenous peoples, Africans, and Europeans.[3] Concern about Brazil's racial makeup can be traced back to the first efforts by the Portuguese to rule their new colony. And ever since the transfer of the Portuguese crown from Lisbon to Rio de Janeiro in 1808, race and racial relations have been central to the political and ideological debates about Brazil's capacity for development and, therefore, the country's destiny. Between 1870 and 1914, the slave-based society collapsed, and the country opened up to European immigration on a massive scale. Politically, Stepan elaborates, this period saw the abolition of the monarchy and the creation of a republic in 1889. Economically, it witnessed Brazil's ever-deepening involvement in the world capitalist system as a supplier of raw materials, such as coffee. Brazil entered the twentieth century a highly stratified society, socially and racially—a society that, though formally a liberal republic, was governed informally by a small, largely white elite and in which, Stepan reminds us, less than 2 percent of the population voted in national elections; the majority of the people were black or mulatto and could not read or write.

Brazilian doubts about the country's racial identity had long been reinforced by racist interpretations of Brazil from abroad. Intellectuals had to contend with the fact that, in European thought, Brazil was held up as a prime example of the "degeneration" that occurred in a racially mixed, tropical nation.

Social medicine of that period, combining racial ideologies with diseases, was particularly concerned about the city, where people of different races and backgrounds lived in close proximity to each other. This concern was linked to a specific scientific theory developed during this period, which explained the illnesses and epidemics of the day by blaming the atmosphere—that is, the quality of the air in Brazil. This air could be poisoned by the putrefaction of organic matter or by emanations from the body, such as sweating. Joel Outtes has argued that the environment itself was placed at the very heart of social reform, thereby incorporating the problem of spatial organization into the reform agenda.[4] Slums and tenement houses in particular were considered dangerous places, "social ills" facilitating the spread of physical as well as moral disease.

Thomas Buckle, Gustave Le Bon, the Count de Gobineau, and other social Darwinists were widely quoted for their theories of Negro inferiority, Mulatto degeneration, and tropical decay.[5] From the United States, the message was the same: as evidence that half-breeds could not produce a sophisticated civilization, anthropologists pointed to Latin Americans, who, they claimed, were now "paying for their racial liberality."[6] Degeneration was seen as a national disease, an ailment that connected individual health to national well-being. The desire to imagine the nation in biological terms, to define in new terms who belonged, was shown in eugenics as the domain of sexuality and race. The notion of sexual gender helped to articulate the notion of race and vice versa, since it was through sexual reproduction that the hereditary component of future generations was modified. Intellectuals looked inward to ask whether they too had a "race spirit" that defined them and gave them a sense of national identity comparable to that of European nations. In comparing themselves to European nations, they often concluded that Brazil needed to develop a true "nationhood" on which they could establish a proper sense of nationalism.

Thus, many of the elites were inclined to think that Latin American countries were not yet proper nations. For them, racial hybridization meant degeneration and regression, and racial differences were biologically determined and could not be changed by social conditions. Policies to encourage European immigration responded to patriotic racial concerns, as voiced by Nina Rodrigues, with "whitening" the national stock, as well as the direct interests of plantation owners in obtaining cheap labor. But, as Dain Borges shows, a wide range of imperial and republican social policies— including the regulation of prostitution; the sanitation of ships, factories, and barracks; the licensing of domestic servants, sports, and physical education; and universal military service—were also justified in terms of protecting the race from contamination or alternatively recuperating its health.[7]

Most liberal elites emphasized the importance of adaptation to environmental factors such as diet, sanitation, and climate.

Procreation was believed to be the agent that permitted any adaptive or "acquired" characteristics. Thus, as Julyan Peard notes, there was hope that through the correct scientific, medical manipulation of the Brazilian environment, a "civilized" people could emerge.[8]

BRANQUEAMENTO

Brazil's small intelligentsia had long been preoccupied with the racial identity and health of the nation. White immigrant labor, it was believed, would contribute to a more progressive society and would improve the country's image as a potentially white nation. The image of the nation as a biological entity whose populations could be "purified" through reproduction deeply affected policies of this era. They began to regulate the flow of peoples across national boundaries and to define in new terms who could belong to the nation and who could not, thereby producing intrusive proposals for new state policies toward individuals.

To a large extent, the educated classes of Latin America shared the racial misgivings of the Europeans. They wished to be white and for this reason, encouragement of European immigration had by the end of the nineteenth century become national policy in Brazil. Representative of this attitude is João Cardoso de Menezes e Souza's *Theses sobre a colonização do Brazil*, published in 1875.[9] It was originally written as a formal report to the Minister of Agriculture on colonization by European immigrants. As Jeffrey Lesser describes, for Menezes e Souza these immigrants were to be the "seed" of municipal life from which would spring the "powerful force of homogeneity and cohesion that will pull together and assimilate" the population at large.[10] A large influx of preferably Northern European people, Menezes e Souza thought, would introduce high culture to Brazil and set the country on the road to modernization. In order to achieve this goal, he would need active propaganda to encourage the stream of immigrants "que nos podem fornecer subsidio de braços laboriosos, para as diversas industrias, sobretudo para a agricola, principal fonte de riqueza no Brazil"[11] (that can supply us with laborious arms, for diverse industries, but above all for agriculture, the main source of wealth in Brazil).

He takes this task as much more than a merely economic endeavor and relates it to the future of the empire. Menezes e Souza must, he so solemnly declares, solve the problems of the past, make an inventory of the country, and reject colonial elements. He compares Brazil to an embryo "em que jaziam embryonarias, as poderosas forças, os inexhauriveis recursos deste gigante da América Meridional"[12] (where powerful forces, the inexhaustible resources of this South American giant, lie like embryos). But because of the unique character of Brazil, he cannot simply follow the examples of other nations because "pouco ou de nada nos serve o exemplo de outros povos de raças, regiões, climas e instituições muito diferentes das do paiz"[13] (the example of other people of very different races, regions, climates, and institutions are of little or no use to us).

Brazil's uniqueness can be interpreted as negative and shameful, a colonial mold, but at the same time there is some essence in the country that should not be lost. Menezes e Souza suggests that leaders, such as he, will have to make an inventory and select which elements will be "good" and which will be "bad" for the future of the country. Although he seems to agree with the racial assumptions of his time period, that Northern European races are superior to Brazil's mixed population, these peoples, their religions, the climate, and their institutions are another unknown but important factor in the future development of Brazil.

Summarizing, Brazil, on the periphery of world capitalism, went through an unsteady and incomplete process of modernization that was contemporaneous with that of Western Europe, with state institutions increasingly assuming authority to control the nation's "health." This period also saw the development of bureaucratic institutions in order to perform public functions. As Thomas Holloway points out, the distrust and often open hostility between the newly institutionalized forces of repression and the sources of resistance in Brazil is related to the imposition of apparently modern bureaucratic institutions of control on a society that was lacking in other fundamental attributes of modernism.[14]

SELF AND OTHER: PROBLEMATICS OF
THE MASTER-SLAVE DICHOTOMY

As Thomas Holloway explains, for obvious reasons, historians of Brazil have often focused on the institution of slavery in their work.[15] One result of this focus, important though such studies have been in expanding our understanding of slavery as a phenomenon and of the slave experience, has been to leave largely unexamined the historical role and experience of another social category that might be generally labeled the "free poor," or the nonslave lower classes. These might include marginal drifters and domestic servants, as well as many people involved in the lower levels of artisan and eventually industrial production, retail trade, and provisioning. In this way, the master-slave dichotomy becomes, in the course of the nineteenth century, less and less adequate for understanding the workings of Brazilian society.

The urban lower class grew both in absolute terms and in proportion to the decline of slavery in the second half of the nineteenth century. These nonslaves were internally diverse and ethnically complex, but in Rio de Janeiro an increasing proportion were Portuguese immigrants. What these urban poor shared in the eyes of the elites were negative attributes: they had neither wealth, status, nor power. To guarantee the presence of "good" civilizing immigrants, it was necessary to weed out the bad ones, those who refused to be exploited, or those who agitated the native populations with "foreign" socialist ideas. Images of the ideal immigrant were constructed not only as white and European but also as hardworking, of a superior moral and cultural background, willing to integrate with local populations, and serious yet able to adapt to a supposedly Brazilian national spirit. All these requirements varied according to the wishes and fears of the elites. It is precisely this multiplicity of images that provided immigrants the space to negotiate their future identity as Brazilians.

Perhaps the most pessimistic determinist, radically opposed to Menendes e Souza, was Raymundo Nina Rodrigues, who persistently argued that the reality of miscegenation, lying beneath the fiction of racial democracy, prevented Brazil from achieving its rightful place among elite nations.[16] Presumably of mulatto origins himself, Nina Rodrigues was born in the conservative and traditionally monarchist state of Maranhão.[17] His preoccupation with the question of race led him to study the determinism of nineteenth-century anthropology and racial "science." Throughout his writing, Nina Rodrigues questions the relationship between race and pathology and, as Julyan Peard points out, consistently disregards human agency as a factor in controlling the progress of the population.[18] Although hygienic living was important, the ramifications of inherited predisposition were more urgent. In the debate over whether hybridization invigorated the race or led to further degeneration, he believed the latter. In the case of crime and the law, he came to argue that "'inferior people' should be granted attenuated" responsibility for any crimes committed, due to their lower intelligence. Those of mixed races would be ranked according to a hierarchy of greater and lesser degeneration and granted civic responsibilities accordingly. He therefore advocated a quasi-segregationist political program, criticizing the smug assumption of Brazilian elites that they understood the mentality of the people over whom they ruled. Much time would have to pass before the Brazilian people could be homogenous. Although frequently quoted, Nina Rodrigues's ideas seem to have had little or no practical impact on the Brazilian criminal law system, and his strict deterministic thought seems directly opposed to that of the Brazilian elites who sought to create or at least to imagine a homogeneous and modern nation.

Another powerful voice against immigration, albeit for very different reasons, is that of Eduardo Prado, a convinced monarchist who published *A ilusão americana* in 1893. The book caused a scandal and was forbidden the day after it was published, probably contributing to its notoriety. In an argument apparently similar to those colleagues from the Hispanic world such as José Martí or José Enrique Rodó, Prado warns Brazil against trusting and befriending the United States, described as an innately corrupt nation, despite its youth:

A podridão é própria dos túmulos e não dos berços. O que há de esperar de uma existência humana cuja infância não tiver sido inocente? . . . o nascer das repúblicas, . . . se não fumegarem em roda do seu berço o incenso puro o a mirra incorruptível do sacrifício o do patriotisimo, não promete e não dará nunca no futuro senão crimes e desgraças.[19]

[Rottenness is a characteristic of tombs, not cribs. What can be expected of a human existence whose infancy was never innocent? . . . the birth of republics . . . if pure incense is not burnt around their crib, or the incorruptible mire of sacrifice and patriotism, promises nothing and in the future will never give anything but crime and misfortune.]

Prado fears the possibility of a military expedition by the Americans in Brazilian territory. As he writes, the United States seemed very interested in the Amazon region at this time:

O General Grant, num discurso pronunciado em 1883, numa recepção ao general mexicano Porfirio Díaz, chegou a dizer que os Estados Unidos necessitavam de tres cousas sómente, porque o resto tudo tinham no seu pais. As tres cousas eram: café, açúcar e borracha. E o general disse: *Seja como for*, havemos de ter café, açúcar e borracha. O general acentuou bem a frase *Seja como for (by any means)*, e no México esta frase foi tomada quase como uma ameaça.[20]

[General Grant, in a speech from 1883, held during a reception for the Mexican general Porfirio Díaz, came to affirm that the United States only needed three things, because his country had everything else. Those three things were coffee, sugar, and rubber. And the general said: By any means, we must have coffee, sugar, and rubber. The general really emphasized the phrase *by any means*, and in Mexico this sentence was almost taken as a threat.]

Hawaii could provide the United States with sugar and Mexico with coffee, but in order to acquire the final commodity, rubber, they needed access to the Amazon, and, in Prado's opinion, the Brazilian government should consider this carefully.

Culturally speaking, Prado also seems to agree with figures such as Martí and Rodó. At heart, Brazil for him is still a colony, failing to have created anything autochthonous that would be comparable to European standards. For Prado, Brazilian elites are "parvenues" who like to show off the biggest and most expensive goods.[21]

Slavery was, of course, an embarrassing topic for the Brazilian elite, since the country had only abolished this institution in 1888, making it the last nation in the Western Hemisphere to do so except for Cuba. Prado prefers not to talk about slavery in his own country but starts out with the treatment of the Chinese by the United States. In the United States, according to Prado, Chinese were occasionally lynched or even burned alive, in spite of a friendship and commercial treaty that the United States blatantly disrespected.[22] Worse still, the Chinese frequently were victims of slavery themselves, such as in the case of Peru:

> Esse tráfico de escravos amarelos era feito por umas casas americanas, e quase sempre sob a bandeira estrelada que protegia a escravidão asiática, já no Peru, já em Cuba. O governo português começou a se impressionar com o escandalo, e o relatório que Eça de Queirós, consul de Portugal na Havana, apresentou ao govêrno demonstrando as monstruosidades cometidas contra os chins, apressou talvez o fechamento do pôrto de Macau à immigração chinesa. Houve americanos estabelecidos no Peru e ligados aos agricultores peruanos que se enfureceram com a suppressão do tráfico amarelo, e foi então que se organizou uma das mais hediondas empresas de pirataria de que há notícia.[23]

> [This traffic of yellow slaves was done by several American businesses; and almost always under the star spangled banner that protected Asian slavery, be it in Peru or Cuba. The Portuguese government was affected by the scandal, and the report that Eça de Queirós, the Portuguese consul in Havanna, presented to the government describing the monstrosities committed against the Chinese, may have hastened the closure of the port city of Macau to Chinese immigration. There were Americans in Peru, connected to Peruvian farmers, who were furious with the suppression of this yellow traffic, and it was then that they organized the most hideous piracy businesses in history.]

Prado tries to defend the prolonged period of Brazilian slavery while attacking the United States at the same time. Whenever a country was about to abolish slavery, the United States opposed its independence, he insists, giving as an example the case of Haiti.[24] After all, Prado reminds his readers, it was England who helped Brazil with its independence, not the United States.[25]

In his attempt to justify slavery in Brazil while condemning the United States, Prado goes as far back as the Greeks and Romans of classical antiquity who had engaged in slavery, which gave them a certain stability and social organization, he claims. Although slavery was never formally abolished, people came to recognize the value of human freedom due to the growth of Christianity. In contrast, in the United States, the end of slavery was characteristically violent, "genuinely American," Prado claims.[26] Compare this to Brazil, he argues, where the country's leaders established a genuinely Brazilian *and* monarchical solution: humanitarian *and* nonviolent. It was clear to Prado that in Brazil, race relations were far superior to those in the United States.[27] While lynchings still took place in the United States, the "Latin" spirit of Brazil always respected freedom and human life.[28] Because of the violent way in which slavery was abolished in the United States, the Americans indirectly helped Brazilians prolong slavery in their own country, which was a major center of slave trafficking as well.[29]

Prado has an equally "original" interpretation of immigration and the accompanying labor issues. A dedicated monarchist, he is fiercely antidemocratic: "O capitalismo semita ou não semita, goza hoje de privilégios reais e efetivos muito mais vexatórios do que os privilégios antigos da nobreza e do clero"[30] (The capitalist system, Semitic or not, today has true and effective privileges that are much more damaging than the old privileges of the nobility and the clergy). In modern life, he elaborates, capital grows by itself, and fate seems to favor only the rich. Republican bourgeois states, such as the United States and France, have no control over this capitalist system. Compare this to the "Roman" and Catholic traditions, where the pope and emperors never lost track of humanitarian ideals, and it becomes obvious that Brazil's Republic is a big mistake.[31] Social problems, labor unrest, and riots, he argues, are worse in the United States than in Europe:

> Grande parte da massa operária é estrangeira, estando ainda na primeira fase da existência do imigrante, fase intermédia, na qual tendo-se desprendido da pátria antiga ainda não adotou a pátria nova . . . Quem não duvidou abandonar a pátria do seu nascimento não tem escrúpulos em perturbar a pátria adotiva. Por

isso . . . o exército operário, nos Estados Unidos, é mais de temer
do que na Europea.[32]

[A large part of the working masses is foreign, still being in the first
phase of the immigrant existence, and in the intermediate phase,
in which they have not yet adopted their host country after having
abandoned their old country . . . Those who abandon their country
of birth have no qualms with causing unrest in their adoptive country. Therefore . . . the workers' army of the United States is much
more threatening than that of Europe.]

The United States is a country of immigrants, who are described
as almost inherent traitors who love neither their country of origin
nor their adoptive country and are thus purely materialistic, search-
ing only for individual gain. Under this framework, they do not
hesitate to foster permanent unrest. In Prado's view, labor agitation
is worse in the United States than in Europe because the American
worker does not have the moral principles and restraints that the
European worker has.

For the author, the worst type of immigrant seems to be embodied by Andrew Carnegie. Carnegie dares to speak of the happiness
of the American worker, a happiness that seemingly spurs Prado
to rage for pages against all things American. Prado argues that
through Carnegie's ruthless exploitation of the foreign born masses,
both he and his followers actually create labor unrest and destroy
any chance of genuine patriotic feelings. Brazil, instead of being
alarmed by this situation, prefers to copy the same disastrous institutions from the United States, institutions that may have even
worse results in Brazil because one cannot simply transplant them
from one country to another.[33]

Despite the censorship of the work, Prado's defense of a national
Latin spirit and the absence of Brazilian racism proved in the long
run to be much more influential than Nina Rodrigues's determinism. Its blatant nationalism and anti-Americanism have overshad-
owed the fear of modernization that is equally forceful in the text.
"Chinese" and "black" slaves are rendered helpless victims without
any means, or even desire, to rebel against their masters. Prado
longs for an intimate Brazilian "national family." What his country
needs, Prado claims, is a father figure, be it a pope, an emperor, or a

traditional Brazilian patriarch, who knows each person individually and can accordingly make just decisions.

RACE AND SPACE: MODERNIZING THE CITY

Holloway describes how the political elite of Brazil felt the need to showcase their successful administration of the state, logically, by promoting Rio de Janeiro.[34] Though Rio de Janeiro was not conceived of as a city-state with the rest of the country as irrelevant periphery, the capital city was at the turn of the century the most important urban center for the local and international elite. Rio, Holloway explains, was in a special position as the capital city, which also meant that its inhabitants were expected to be on their best behavior; the need for public tranquility went beyond the requirements of the local business community. Contributing to this order was the institutional and physical proximity between the highest levels of the national government and life on the streets.

The underlying rationale for this project was not only to make the urban population more "civilized" and therefore more likely fit European paradigms of development but also to increase control of the population through state structures, filling the gap in the mechanisms of domination and subordination left by the relative decline of personalized hierarchies of patronage and power that had served in earlier times. As Holloway correctly observes, however, private relationships of domination and subordination never faded away completely.[35] The development of impersonal state institutions to fill the gap in the public sphere served to relieve the strain on the personalistic system of control, thus allowing it to thrive. In this sense, there was no confrontation between modernizing institutions, such as the encouragement of European immigration, and the traditional social hierarchy. It seems that one actually complemented the other. Brazil's partial or inconclusive transition to modernity, then, helped the country cope with growth and change since it was managed according to the interests of the small group of people who were the heirs of the colonial elite.

During the first two decades of the twentieth century, the idea of city planning, defined as a project that took the whole city as a site of intervention, was established in both Brazil and Argentina. City planning of that period is striking for its racial prejudice and fear of the masses. In 1906, Argentine public officials pro-claimed the need for a plan that would modernize and sanitize Buenos Aires, and in Brazil, engineer and architect Victor da Silva Freire talked about this same need.[36] After the second half of the nineteenth century, as Botafogo, Tijuca, and other more outlying areas became accessible to the middle class as residential enclaves, many old downtown buildings were converted into tenements known formally as *estalagens* and informally as *cortiços*, or bee-hives—inner-city slum dwellings where people lived in extremely poor and unhygienic circumstances.[37] By police count, accord-ing to Holloway, by 1875 some 33,000 people—more than 10 percent of the downtown population—lived in *cortiços* and many more in *cabeças de porco*, rooming houses, recalling the maze of cavities left after boiling down the skull of a pig for headcheese. (The name of the rival *cortiço* at the end of Azevedo's novel that is discussed later, *Cabeça de gato*, or cat's head, is an ironic indicator of its tenants' desperate poverty.)

The decline of slavery at this time facilitated this geographic relocation, in part because the gradual transition to free labor relieved members of the upper class of their burden of continually having to control the working population (previously their own property). Nonslave workers who could afford neither suburban housing nor the attendant transportation costs began to concen-trate in the downtown area, and the state-sponsored apparatus of repression moved in to replace the control mechanisms that mem-bers of the former slave-owning class increasingly abdicated as they moved with their houseservants to the suburbs.[38]

BRAZILIAN LAWLESSNESS: *O CORTIÇO*

The naturalist novelists inhabited the city as both voyeur and active participant, and their writing comprised an act of creation—not just of literary art but, in certain ways, of Carioca (as those from Rio de Janeiro are called) urban society itself. *O cortiço*, by Aluísio de Azevedo, is a text that has almost exclusively been seen as a novel about class conflict. Critics have tended to emphasize the simi-larities between Azevedo and Emile Zola. While Latin American writers without any doubt had a strong connection to French natu-ralism, I prefer to think of the Latin American naturalist novel as appropriating existing models for its own unique ends rather than seeing them as mere pale imitations of European counterparts. By juxtaposing *O cortiço* with the sociological texts of Nina Rodrigues, Prado, and Souza, I suggest the novel can also be read as a reaction against the positivist "whitening" ideologies that stimulated European immigration. While the miserable conditions of Rio de Janeiro's poor are described in great detail, issues of racial politics, equally important to the plot, have generally been neglected by literary critics. It is surprising how this novel has barely been read for its racial matter.

O cortiço, published in 1890 and translated into English first as *A Brazilian Tenement* in 1926 and more recently as *The Slum* in 2000, deals with three Portuguese immigrants in Rio de Janeiro in the 1870s and their adaptation to Brazilian society. At the same time, it is a reflection of the social ills that plagued the Carioca society of Azevedo's time, the rapid modernization of Rio de Janeiro, and an examination of the uneasy cohabitation of recent European immigrants and slaves. The *cortiço* is converted into the public space for the poor, where the cultural transactions between agricultural tradition, immigrant backgrounds, and a growing urban modernity take place.

Still surprisingly readable for a twenty-first-century audience, *O cortiço* is a deeply pessimistic reflection of the social ills that plagued the Carioca society of the 1870s, the period immediately preceding abolition. The novel follows the lives of Portuguese immigrants as they adapt to their new country. A reading of *O cortiço* from a Foucauldian perspective is particularly fruitful since Azevedo's work is often read as simply describing "reality." Ethnographer and sociologist Gilberto Freyre, for instance, affirms, "Deixou Aluísio Azevedo no seu *Cortiço* um retrato disfarçado em romance que é menos ficção literária que documentação sociológica de uma fase e de um aspecto característico da formação brasileira"[39] (In his novel *O cortiço*, Aluísio Azevedo left a portrait disguised as fiction. It is less literary fiction than sociological documentation of a phase and a characteristic aspect of the formation of Brazil).

Freyre's admiration for the novel is not surprising, given Azevedo's approach to his subject matter. According to Artur Azevedo, his brother's "scientific" attitude was completely new in Brazilian letters at that time: "Os brasileiros que até hoje se têm esgrimido no romance. . . . escolheram sempre uma sociedade convencional . . . Aluísio Azevedo foi aos cortiços, metóse entre essa população heterogênea das estalagens"[40] (Brazilians who up to this day have written novels . . . always chose conventional society . . . Aluísio Azevedo went to the *cortiços*, he interacted with this heterogeneous group of people in the tenement buildings). A friend recalls how Azevedo disguised himself in order to do his "fieldwork": "Os primeiros apontamentos para *O cortiço* foram colhidos em minha companhia em 1884, numas excursões para 'estudar costumes,' nas quais saímos disfarçados com vestimenta popular: tamanco sem meia, velhas calças . . . camisas de meia rotas nos cotovelos . . ."[41] (The first notes for *O cortiço* were taken in my company in 1884, on excursions to "study customs," in which we walked around disguised in rustic clothes: clogs without stockings, old trousers . . . shirts torn at the elbows . . .)

The free poor of nineteenth-century Rio lived in a social world ruled by the commercial bourgeoisie and a political elite who had inherited the colonial project on the periphery of the capitalist world economy. In their eager embrace of the neocolonial order, the elite needed to deal with the social menace to that order as cities like Rio de Janeiro grew and changed. This menace needed to be classified and analyzed in order to find an appropriate "cure." Azevedo disguises himself as if he were a spy in order to gain access to this "other" Rio. Amy Chazkel calls this methodology of collecting stories and entering unfamiliar territory such as tenement blocks in the line of duty a kind of proto-ethnography.[42] The patio is converted into the public space where the cultural transactions between agricultural tradition, immigrant backgrounds, and a growing urban modernity take place. In journalistic descriptions of the late nineteenth century, the *cortiço* appears as a foreign cyst in the body of the supposedly civilizing city, as is evident in the article "Saúde pública e limpeza da cidade," which appeared in *Gazeta de notícias* on June 18, 1876:

> O estudo da vida nos cortiços e a estatística de seus habitantes, daria assunto por si só para largas observações. Dentro desta cidade em que estamos, há outras pequenas cidades que ninguém vê . . . No meio de uma quadra de casas, há um pequeno portão, com um

longo corredor, e no fim um pequeno pátio circundado de verda-
deiros pombais onde vive uma população. É aí o cortiço.[43]

[The study of life in the *cortiços* and the statistics of its inhabitants
is itself a topic that allows for long periods of observation. Within
this city in which we live, there are other small cities that nobody
sees . . . In the middle of a block of houses, there is a small door,
with a long corridor, and at the end a small courtyard surrounded
by tiny shacks where people live. This is the *cortiço*.]

The slum-dwelling urban poor were described by the city plan-
ners as degenerate, criminal individuals, and this depiction of the
urban poor served to justify government intervention in such areas
in order to improve them through the construction of a newly
constructed environment in accordance with the needs of capital
accumulation.[44] The group that most agitated reformers in Brazil
was largely black and mulatto. The reformist professionals assumed
that social ills accumulated at the bottom of the racial and social
hierarchies, that the poor were poor because they were unhygienic,
dirty, ignorant, and genetically unfit. Frequently, the "problems"
caused by the *cortiço* made the news. These dwellings were seen as
fostering a legitimate health risk, and the slum dweller was thought
to endanger the population as a whole: "[U]dos focos de infecção
do Rio de Janeiro eram os cortiços . . . esses focos de miasmas,
sacrificando-se a saúde pública"[45] (One of the places of infection in
Rio de Janeiro was the *cortiços* . . . those centers of miasmas, com-
promise public health).

The particular "races" that concerned these reformers were not
preexisting, discrete, biological entities, but rather sociopolitical
categories created through scientific work and the social relations
of power. Thus, closely related to the process of racial whitening,
city planning was linked to a project for modernizing and improving the race. This explains why the rise of city planning took place
during the same period in which government began to actively
intervene in social questions of racial improvement.

The Rise of the Poor White Immigrant: Stereotyping the Portuguese

After abolition, former slaves were left without education or compensation and were often forced to compete on unfavorable terms for wage labor with the more than 1.5 million white immigrants who entered the country between 1890 and 1920. With the republic, in which all were supposed to be equal, the notion of citizenship became crucial. The perfect citizens would be those who accumulate wealth through hard work and dedication. In this way, work starts to dignify the citizens, qualifying them and legitimizing their riches. Work becomes a means through which the immigrant can negotiate his or her new Brazilian identity. As Gladys Ribeiro asserts, "O trabalho assalariado foi introduzido no Brasil, e a figura do trabalhador livre, nas condições específicas do processo histórico de formação do mercado de mão-de-obra entre nós, provocou um novo entendimento de como se originava o lucro"[46] (Salaried labor was introduced in Brazil, and the figure of the free worker, under the specific conditions of the historical process of the formation of a national market for labor, led to a new understanding of how wealth could be created). The presence of poor white immigrants in the cities was disturbing to those who wanted to "civilize" Brazil. What was previously thought to be nonexistent became a reality: the existence of poor Europeans, whose social conditions were barely distinguishable from those of blacks, personifying forms of social decadence that had previously been attributed exclusively to blacks. These poor, obviously white, Portuguese immigrants were being turned into ugly crooks with no moral scruples, and from that point one need take only a small step to see them as being "black." Interestingly enough, these immigrants, all from the Minho region in Northern Portugal, were occasionally compared to African slaves in the media. In 1870, the newspaper *O Povo*, for example, published an article titled "Paralelo entre africanos e portugueses," in which the author affirms the following: "Se diferença se pode dar no seu físico, certo que na moralidade das ações, muitas vezes o africano excede ao português . . . O português que para aqui vem é réu de polícia, ladrão de estrada, chefe de quadrilhas, passador de papel falso, galegos que correspondem ao que chamamos negro cangueiro"[47] (Even if one can see the difference in their physical appearance, it is evident that in the morality of their actions, often the African is superior . . . the Portuguese who come here are police suspects, street thugs, gang leaders, counterfeiters, brutes who deserve we call them lazy blacks).

According to Ribeiro, between 1830 and 1930 about one million Portuguese immigrated to Brazil.[48] The concept of a good immigrant also varied within Brazil. Ribeiro points out that whereas in São Paulo agricultural colonies were formed to help with the harvest, Rio's immigration was much more urban in character, with immigrants working in businesses where they were expected to be diligent and respectful. All were of extremely poor origin, and mutual solidarity was the best defense to help them survive in the new country.[49]

Due to their supposedly innate conservatism, the Portuguese were seen as defenders of the status quo—that is, as enemies of the new republic. Interestingly enough, they were also accused of causing political unrest among Rio's poor because of their exploitation of other immigrant groups. The Portuguese, though culturally inferior, could be considered both white and good immigrants because of this supposedly desirable work ethic. Ribeiro notes the following:

> Pelo seu trabalho assíduo e árduo, o português as vezes tornou-se rico . . . Haveria algo melhor do que isto, em um Rio de Janeiro que se reestruturava, caminhando com firmeza para um projeto sócio-econômico nitidamente capitalista? È final do século XIX no Brasil. Levas e levas de imigrantes desembarcam em nossos portos, expulsos do campo europeu pelo avanço das relações de produção capitalistas na Europa. Chegados aqui, passam a substituir o trabalhador escravo no campo e na cidade . . . Pelos discursos da época, o Brasil estaria entrando em uma nova fase da sua história . . . Seria um país comparável às grandes nações européias, civilizadas e modernas. A cidade do Rio de Janeiro seria a vitrina das transformações que nos levariam ao progresso.[50]

[For their assiduous and arduous work, the Portuguese sometimes become rich . . . Could there be anything better than this, in a Rio de Janeiro that is restructuring itself, firmly treading toward a clearly capitalistic socioeconomic project? It is the end of the nineteenth century in Brazil. Group after group of immigrants disembark in our ports, cast from European fields by the advance of capitalistic relationships of production in Europe. Upon arrival, they come to substitute the slave laborer in the country and the city . . . For the discourses of the era, Brazil should be entering a new phase in its history . . . It should be a country comparable to the great European nations, civilized and modern. The city of Rio de Janeiro should be the display window of the transformations that will carry us to progress.]

Ribeiro focuses on images of Portuguese from newspapers and criminal trials and makes a convincing argument that anti-Portuguese sentiments in Rio are connected to the resistance to this disciplined market economy. The plurality of conceptions of autonomy, freedom, and work ethic generated disputes, rivalries, and conflicts among Rio's working classes. As the largest immigrant group, they replaced the slaves in the workplace. She notes that it took many years for both Portuguese and Brazilian-born Portuguese to incorporate Brazilian culture. The Portuguese who came to Brazil by the late nineteenth century stayed in urban centers, mainly Rio de Janeiro. This immigration coincided with the abolition of slavery in 1888 and the proclamation of the republic.[51] These events inspired anti-Portuguese sentiments. Instead of being seen as explorers and colonizers, these new Portuguese became increasingly known as "foreigners," monarchists, and conspirators against the republic. Those who owned import or distribution businesses most often did not opt to attain citizenship. These businesses were run in a strictly hierarchical way, with the lowest jobs filled by recent immigrants. This type of business remained dependent on the metropolitan financial and mercantile center, which was usually located in Porto, and the workers came mostly from the Minho region in Northern Portugal. This resistance toward assimilation obviously provoked resentment among Brazilians.

IMPOSSIBLE CITIZENS

The failure of the institutions that were supposed to educate and protect these new citizens provides the framework for *O cortiço*. What kind of country welcomes new citizens when very few people would be able to provide models for appropriate behavior? How can immigrants possibly improve their new country if all the local institutions are set on destroying whatever modernizing effects immigration could have? What will be the result of this mingling of white immigrants with free blacks and mulattos? Unlike Prado, labor unrest seems hardly a problem for Azevedo. There are a few minor "foreign" characters in the novel, but their contribution to progress seems insignificant. For Azevedo, the immigrant is slightly ridiculous and inclined to participate in corrupt Carioca society. For example, a group of Italians who work in a pasta factory are consistently compared to parrots, and there is a brief mention of a Chinese man peddling shrimp who finds himself unwittingly involved in a fight and is subsequently thrown out of the *cortiço*: "Era o que faltava que viesse também aquêle salamaleque do inferno para azoinar uma criatura mais do que já estava!"[52] (Just what we needed: Some goddamned Chink to kick up even more of a fuss!) In a clear nod to French naturalism (anti-Semitism was not an issue in Brazil at the time), there is an ugly and stingy Jew, Libório, who is tolerated and even fed by the tenants of the *cortiço* but fails to do anything in return: "Na estalagem diziam todavia que Libório tinha dinheiro aferrolhado, contra o que ele protestava ressentido, jurando a sua extrema penaria"[53] (When people at São Romão claimed he had money hidden away, he protested indignantly, swearing that he hadn't a penny). While João Romão, the founder and owner of the *cortiço*, made his fortune by wisely investing his savings, Libório's (supposedly typically Semitic) capitalist strategy of saving and scraping by, then, turns out to be unproductive. It is not until João Romão finds Libório's money and discovers that a large part of it is no longer valid: "[S]ofreu uma dolorosa decepção: quase todas as cédulas estavam já prescritas pelo Tesouro; veio-lhe então o receio de que a melhor parte do bolo se achasse inutilizada"[54] (He suffered a painful disappointment: the bills were so old that they could no longer be redeemed, and he began to worry that perhaps most of his jackpot would prove worthless). He decides to use the bills as counterfeit money in an attempt to finance the rebuilding of the *cortiço*. This is the only part of Libório's legacy that has any impact on Brazilian society.

Taking a closer look at *O cortiço*, it becomes clear that the division between "black" and "white" is by no means simple. The different versions of "whiteness" are represented by the three main Portuguese immigrant characters. The first immigrant, Miranda, enjoys some economic success at the beginning of the novel—largely because of the dowry of his Brazilian-born wife, Estela—but in return he has to tolerate her numerous infidelities. He moves to the neighborhood of Botofago, escaping the increasingly unsafe shady downtown and hopes to establish a more aristocratic lifestyle. It is made clear from the beginning that Miranda is a decadent individual and will not help invigorate his adopted country in any way: "A mulher, Dona Estela, senhora pretensiosa e com fumaças de nobreza, já não podia suportar a residência no centro da cidade como também sua menina, a Zulmirinha, crescia muito pálida e precisava de largueza para enrijar e tomar corpo"[55] (His wife, Dona Estela, a pretentious lady of aristocratic airs, could no longer endure life in the center of town, and his daughter, Zulmira, was unnaturally pale and needed fresh air to fill out and grow stronger). To fill his empty, meaningless life, Miranda buys himself the title of baron and pretends to be an aristocrat. Immigrants such as he will not change Brazil. He literally buys himself respectability in order to enter into a social class to which he has no right to belong. This is a corrupt and unproductive social class; he transforms himself into a Brazilian that will not make any major contributions to modernizing the country.

In contrast, the second immigrant, João Romão, is an extremely hard worker and a shrewd businessman, amassing his fortune by starting a small restaurant and by cheating his customers. Next to Miranda's pretentious house, he constructs his own *cortiço*, stealing materials and tools, exploiting workers, and renting houses to lower-income families and washtubs for laundresses. He continues to change and progress to a much larger extent than Miranda. But João Romão too will eventually become "Brazilianized" when he imitates Brazil's cruel race relations and makes a slave, Bertoleza, erroneously believe that she has bought her freedom from him (the novel is set in the 1870s, before the official abolition of slavery in 1888). Having won her trust, he is given her wages to invest and he lends the money against high interests while keeping the profits for himself. Bertoleza also becomes his concubine, cooking and cleaning for him for free. Because he is white and she is black, she looks for white men, "como toda a cafuza"[56] (like all colored women). He achieves his wealth through the exploitation of the black population, just as Brazilians had always done. Even João Romão, who never washes or changes his clothes and wears wooden clogs instead of shoes, will eventually learn table manners and dress himself appropriately in order to marry Miranda's daughter. In a very cynical comment by Azevedo about the relations between immigrants and slaves, João Romão eventually rids himself of Bertoleza by giving her over to the authorities, claiming to simply be returning this runaway slave to her rightful master. João Romão may not truly believe in racial inferiority, but he is described as being consumed by avarice, to the point of insanity, considered at that time to be a typical immigrant's disease: "Aquilo já não era ambição, era uma molestia nervosa, uma loucura, um desespêro de acumular, de reduzir tudo a moeda"[57] (It had gone beyond ambition and became a nervous disorder, a form of lunacy, an obsessive need to turn everything into cash).

Immigrants like João Romão came to Brazil to attain material wealth and therefore seem to be interested neither in politics nor in the general well-being of their adoptive country. Avarice corrupts, and where it abounds, there is no room for patriotism. This moral flaw is at once the cause and the effect of his behavior produced by Carioca society, since it fails to uncover the immigrant's dishonesty, as much as corrupting it, by his building of the *cortiço*. Carioca society fails to expose the immigrant's dishonesty, both in his treatment of Bertoleza and in his illegal building of the *cortiço*. The beginning of the *cortiço* is clearly clandestine. Not only does João Romão appropriate the savings and labor of the supposedly freed slave, Bertoleza, but also all the construction is done with stolen materials:

Que milagres de esperteza e de economia não realizou ele nessa construção! Servia de pedreiro, amassava e carregava barro, quebrava pedra; pedra, que o velhaco, fora de horas, junto com a amiga, furtavam à pedreira do fundo, da mesma forma que subtraiam o material das casas em obra que havia por ali perto . . . Nada lhes escapava, nem mesmo as escadas dos pedreiros, os cavalos de pau, o banco ou a ferramenta dos marceneiros. E o fato é que aquelas três casinhas, tão engenhosamente construídas, foram o ponto de partida do grande cortiço de São Romão.[58]

[What prodigies of cunning and frugality he realized in their construction! He was his own bricklayer, he mixed and carried mortar, he cut the stone himself—stone he and Bertoleza stole from the quarry at night, just as he robbed all the nearby construction sites . . . (They) took everything, including bricklayers' ladders, sawhorses, benches, and carpenters' tools. And the fact was that those three two-room houses, so ingeniously constructed, were the point of departure for a huge slum later dubbed São Romão.]

The third Portuguese, Jerônimo, is apparently a model immigrant. He is an honest stonecutter, who lives in the tenement and works for João Romão. He is, at least initially, a virtuous family man, devoted to his wife, Piedade. The couple tries to maintain their rural Portuguese culture and resist assimilation by being extremely hardworking and virtuous, listening to *fado*, traditional Portuguese songs, eating Portuguese rather than Brazilian food, and having a less rigorous concept of cleanliness. As a stonecutter, Jerônimo tries to tame Rio's unbridled nature, breaking down its mountains for granite that will be used in the new buildings of the quickly expanding city. As he is looking for a job in the granite quarry, he describes the people working there: "Aquêles homens gotejantes de suor, bêbedos de calor, desvairados de insolação, a quebrarem, a espicaçarem, a torturarem a pedram pareciam um punhado de demónios revoltados na sua impotência contra o impassível gigante que os contemplava com desprêzo, imperturbável"[59] (Those men dripping sweat, drunk with the heat, crazed with sunstroke, pounding, stabbing, and tormenting the stone seemed a band of puny devils rising up against an impassive giant who scornfully looked down at them, indifferent). In this titanic struggle between nature and progress, Jerônimo, however, is not impressed by nature's force. He examines the quarry, explaining to his future employer how profits could be greatly increased: "O membrudo cacouqueiro havia chegado à fralda do orgulhoso monstro de pedra; tinha-o cara a cara, mediu-o de alto a baixo, arrogante, num desafio surdo"[60] (The mighty Portuguese stonecutter reached the foot of that haughty stone monster. Standing face-to-face with it, he took its measure silently and defiantly).

Nevertheless, this fecund union between nature and progress quickly deteriorates as Jerônimo falls for his neighbor, the beautiful, free-spirited Rita, a mulatta from Bahia, and abandons his family for her. He meets her in the courtyard, where he was playing fados on his guitar and all of a sudden hears music from Bahia and sees her dancing: "Naquela mulata estava o grande mistério, a síntese das impressões que ele recebeu chegando aqui: ela era a luz ardente do meio-dia . . . era o veneno e era o açúcar gostoso"[61] (That mulatta embodied the mystery, the synthesis of everything he had experienced since his arrival in Brazil. She was the blazing light of midday . . . She was poison and sugar). Rita, through her mixed African and Portuguese heritage, seems to embody the spirit of Brazil, sweet but dangerous, always seductive, "volúvel como toda a mestiça"[62] (mercurial like all half-breed women). To prove his love for Rita, Jerônimo murders his rival and runs away with this new girl, abandoning his family. His process of becoming a Brazilian is described as a complete transformation: "Adquiria desejos, tomava gôsto aos prazeres, a volvia-se preguiçoso resignando-se, vencido, às imposicões do sol e do calor, muralha de fogo com que o espírito eternamente revoltado do último tambor entrincheirou a pátria contra os conquistadores aventureiros"[63] (He developed desires, enjoyed his pleasures, and grew lazy, bowing in defeat before the blazing sun and hot weather, a wall of fire behind which the last Tamoio Indian's rebellious spirit defends its fatherland against conquering adventurers from overseas).

Brazil's sun and climate are capable of utterly destroying immigrants and their ambitions, as if its nature were a curse against foreign invaders. Brazil seems to be governed by nature's primitive impulses rather than by strong civilized governmental institutions. The police are inefficient, as is shown in the episode of Florinda's rape, or simply corrupt, as when they trash and steal everything in the *cortiço* when a house in the *cortiço* is on fire. In addition, Alexandre, a policeman who lives in the *cortiço*, never tries to impose any sense of order. Private institutions do not help to control these lawless people either. Promiscuity in the *cortiço* is notorious; such is the case of Leocádia, who wants to become pregnant in order to get a job as a wet nurse, and the rape of Florinda, who is later thrown out by her mother. Neither the police nor the property manager, in this case, João Romão, does a thing to help.

The traditional family seems increasingly at risk with the growing presence of women in the labor force; the new sexual mores that come with modernity; as well as the immigration, prostitution, illegitimacy, illegal abortions, and alcoholism that accompanied increasing industrialization, internal migration, and urbanization. All three immigrants fail to establish a type of family life that is suitable for the growth of the nation. Miranda incorrectly suspects that his daughter is not his and is unable to feel any fatherly feelings for her. João Romão marries this possible bastard child in spite of his having lived with Bertoleza, a black slave who believes herself to be freed but becomes, in fact, a victim of fraud when João Romão forges an official document declaring her free. He can escape the union with the betrayed slave by using the very same laws that are supposed to protect and stabilize the country. And Piedade, the one legitimate spouse, becomes a prostitute and an alcoholic in this cruel Brazilian society when Jerônimo abandons her in favor of his mistress.

The Modern Woman

While Azevedo mostly rejects determinist notions of race in his portrayal of male characters, his attitude toward women, particularly immigrant women, is more ambiguous. I agree with Elizabeth Marchant that Azevedo shows a peculiarly hostile attitude toward his female characters in general,[64] though I feel that this reliance on racial determinism when dealing with female characters is by no means unique to Azevedo but is deeply tied to the rhetoric of modernization itself.

If the experience of modernity brought with it a sense of innovation and chaotic change, it simultaneously engendered multiple expressions of desire for stability and continuity. Nostalgia, such as mourning for an idealized past, emerges as a formative theme of modernity. In this sense, the age of progress was also the age of yearning for an imaginary lost paradise. This desire to regress, as Rita Felski correctly points out, was attributed to the dislocations of the modern age, as increased mobility and demographic shifts caused significant sectors of the population to be uprooted from their native lands and hence to lose a sense of continuity with their birthplace and its history.[65] Located within the household and an intimate web of familial relations, more closely linked to nature through their reproductive capacity, women occupied a sphere of timeless authenticity seemingly untouched by the alienation and fragmentation of modern life. By virtue of their position outside the dehumanizing structures of the capitalist economy as well as the rigorous demands of public life, Felski asserts, the woman became a symbol of inalienable, and hence unmodern, identity.

Anne McClintock observes that the inherent ambivalence in nationalisms—veering between nostalgia for the past, and the impatient, progressive discarding of that same past—is typically resolved by figuring this contradiction as a "natural" division of gender.[66] Women are represented as the atavistic and authentic site of national tradition, inert, backward-looking, and natural, thus embodying nationalism's conservative principle of continuity. Men, by contrast, represent the agent of national modernity: forward-thrusting, potent, and historic, embodying nationalism's progressive principle of discontinuity. The equation of woman with nature and tradition, already commonplace in early modern thought, received a new impetus from the popularity of Darwinian models of evolutionary development, residing in an explicit contrast between a striving restless masculinity and an organic, undifferentiated femininity.[67]

As mothers, women were often seen as objects of nostalgia but rarely as the subject of it. Rather than desiring the past, they *were* the past: intrinsically linked to the domestic sphere. The adulteress Estela suggests practicing a proscribed sexuality that corrupts the social function of motherhood and, therefore, of the family and founding institute of the nation. What happens then if she immigrates? The immigrant woman is inevitably trapped; as a female, she is assumed to be unchanging, authentic, but her immigrant status makes her a suspicious receptacle of any national tradition; she might assimilate too rapidly. She is supposed to represent both continuity, as a female, and rupture and modernity, as an immigrant. She thus often seems trapped outside the law, caught between insanity and prostitution.

There are two immigrant women in the novel. The first is the expensive French prostitute Léonie, whose beauty is totally artificial and who is interested only in extracting money from her suitors. This woman has become the embodiment of the antinatural and is solely identified with money, since both money and woman are capable of contaminating. This prostitute is seen as the tyrannical symbol of an unbridled female sexuality linked to contamination, disease, and the breakdown of social hierarchies in the modern city. The prostitute is an insistently visible reminder of the potential anonymity of women in the modern city and shows the separation of sexuality from familial and communal bonds. Léonie is all artifice, appearance without substance. Like many of the prostitutes portrayed by Émile Zola, Leónie has sexual relationships with women as well as with men, and she rapes a young girl in the tenement, lures her into prostitution, and will eventually offer to take care of Jerônimo and Piedade's daughter. As a corrupter of family and society at large, Léonie embodies the worst fears and prejudices that surround immigrant women. Piedade clearly does not adapt to Brazil and is shown to be homesick for Portugal. While Azevedo seems to reject deterministic conceptions of race in his male characters, his female characters on the other hand do seem inevitably to follow the path of their biological disposition. Throughout the novel, we see a strong hostility toward female immigrant characters. While Jerônimo adapts himself to his new country, his wife, Piedade, cannot. She claims to be made of one solid block (nationality), and to change her would imply breaking her apart. Piedade represents a more traditional type of immigration, in which one was supposed to return to Portugal after achieving financial success in Brazil. As essential creatures, women apparently do not have the same possibilities of reinventing themselves in a new country as men do. When confronted with change, they fall into the corrupting forces of prostitution or insanity, as is the case of Piedade, who, in addition to going insane, is animalized as well. This process begins when her husband notices her rancid odor caused by lack of hygiene and his sudden distaste for her Portuguese codfish with potatoes and boiled onions, and it ends with her fall into alcoholism, prostitution, and insanity.

In spite of its critique of Carioca modernity, *O cortiço* never idealizes premodern society, like Eduardo Prado or Gilberto Freyre. For Azevedo, the provincial writer (he was born in São João de Maranhão, a small provincial capital in the rural Northeast) and tireless abolitionist, the legacy of Brazil's past was not a Golden Age but a curse and embarrassment. For Azevedo, European-born immigrants do not have any positive impact on Carioca society. Brazil is simply too overwhelming, and the Europeans who come in the hope of making it in America discover that they are being remade by this new, out-of-control country. For Azevedo, Brazil will never become a modern country unless it addresses its heritage of slavery. Immigrants of whatever race will not contribute to improving their new country if the local institutions of slavery and the pseudo-aristocratic culture are determined to destroy whatever modernizing effects immigration could possibly have. Immigrants are seduced as irrevocably and fatally as Jerônimo is seduced by Rita. Thus, Azevedo attacks the ideologies of *branqueamento* by pointing out that becoming "Brazilian," rather than a privilege bestowed on the few, is an inevitable curse.

Surprisingly, Azevedo was employed as a diplomat by the Baron of Rio Branco, the Minister of Foreign Affairs from 1902 to 1912, who used a group of famous "European-looking" writers to improve Brazil's image in hopes of stimulating European immigration. While Rio Branco and Azevedo were both fierce abolitionists (the "Lei do Ventre Livre" [Law of the Free Womb] proposed by Rio Branco and passed in 1871 is generally seen as an important step toward abolition), Rio Branco was a strong believer in the benefits of European immigration. I have been unable to find a clear reason for this apparent change in Azevedo's thought. Azevedo's biographer, Jean-Yves Mérian, apparently does not see any contradiction in Azevedo's pessimism concerning Brazil and this elite group of diplomats who were supposed to promote the country, and he leaves this drastic change in Azevedo's thinking unexamined.

AN EXAMPLE WORTH FOLLOWING: ANGLOPHILIA
IN *A CASA VERDE* BY JÚLIA LOPES DE ALMEIDA

Unlike Azevedo, whose fame and reputation never diminished in Brazilian literature, Júlia Lopes de Almeida (1863–1934) has only been rediscovered by feminist literary scholars who started republishing her works in the 1980s through the publisher Editorial de Mulheres.[68] However, at the peak of her fame, she was a highly respected novelist and essayist, who, according to her husband, the Portuguese poet Felinto de Almeida, was not included into the Brazilian Academy of Letters because of her gender.[69] Mary Louise Pratt has argued that discourses by nineteenth-century female public intellectuals should be read through the theoretical prism of Latin American racial positivist ideologies, in which female subordination was justified because of women's supposedly biological and intellectual inferiority.[70] Although, as Pratt argues, various lettered women were widely read and respected in their own time, they always had to speak as women, never directly defying male privileges.

A daughter of Portuguese immigrants, throughout her career, Júlia Lopes de Almeida was continually looking for new paths that women could follow in order to help Brazil become a "modern" and "civilized" country. Rejecting deterministic theories on the inferiority of women, Lopes de Almeida throughout her work always insist on the correlation between the ills of society and the crisis of gender roles. Almeida was a genius in combining the image of the dutiful bourgeois housewife with that of a professional writer who was much better known and respected than her husband. She cleverly never addressed "male" issues but preferred to speak to a supposedly female audience, whether in her conduct manuals, essays, or novels. In the interview titled "Um lar de artistas" ("An Artist's Home"), conducted by journalist João do Rio, Lopes de Almeida describes in great detail the atmosphere of a happy bourgeois family home, living in "um *cottage* admirável, construído entre as árvores seculares da estrada de Santa Teresa"[71] (an admirable cottage, constructed amidst the secular trees of Santa Teresa Avenue). Although the journalist is full of admiration for the author's work, Lopes de Almeida presents herself, first and foremost, as a dutiful wife and mother: "Mas não há meio de esquecer a casa. Ora entra uma criada a fazer perguntas, ora é uma das crianças que chora. Às vezes não posso absolutamente sentar-me cinco minutos"[72] (But there is no way one should neglect the home. Now a servant comes to ask you questions, now one of the children is crying. Sometimes I cannot even sit down for five minutes).

In the same interview, she mentions that her favorite novel is *A casa verde* "porque foi escrito em colaboração com meu marido. *A Casa Verde* lembra-me uma porção de momentos felizes"[73] (because it was written in collaboration with my husband. *A casa verde* reminds me of many happy moments). Few critics would agree with this assessment, however. Some of the most interesting novels are *Memórias de Marta* (*Marta's Memories*, 1889, reprinted in 2008), about a young girl growing up in a *cortiço* in Rio de Janeiro; *A intrusa* (*The Intruder*, 1908, reprinted in 1994), about a much maligned governess; *A falência* (*The Bankruptcy*, 1901, reprinted in 1978), about the bankruptcy and suicide of a wealthy businessman and the impact on his family; *A família Medeiros* (*The Medeiros Family*, 1892), an abolitionist novel; and *A viúva Simões* (*The Widow Simões*, 1897, reprinted in 1997), which tells of a love affair of a young widow who was never able to love her husband. All of these works deserve a much wider readership and should be translated. By comparison, *A casa verde* is clearly a minor work, apparently written in great haste and intended simply to earn some money for the quickly diminishing finances of the author's family. *A casa verde* was published as a book in 1932, although it was written in weekly installments by both Lopes de Almeida and her husband under the pseudonym A. Julinto for the newspaper *Jornal do Comércio* in 1898 and 1899. By 1932, however, literary tastes had changed, and Francisco Alves, her previous publisher, was no longer interested in her oeuvre. The family had fallen on hard times, and Almeida published the novel herself, unsuccessfully trying to turn the tide and earn some money.[74] In many ways, *A casa verde* seems to be all that Almeida despised, a "*romance de folletim*" (a feuilleton, or serial romance novel) with spectacular plot twists, a female protagonist that disobeys her father, exotic gypsies, and an outrageous villain. Almeida's pecuniary difficulties and the collaboration with her husband, who was a poet and not a novelist,

probably explain the lack of literary quality of the work as well as
her betrayal of her literary convictions. However, I hope to show
that a closer reading reveals that the clashing discourses and the
sloppy, "unfinished" plot highlight contradictions in racial and
gender assumptions of the fin-de-siècle Brazil.

A constant theme in all of Almeida's writings is the poor education women receive within a patriarchal family, thus resulting
in the inability of the Brazilian woman to be a proper wife and
mother of a healthy family.[75] Almeida wanted to transform these
frivolous and useless women into domestic partners that would
be able to help their husbands in times of need or, indeed, be
able to work for a living if required. Writing against the stereo-
type of the indolent, ignorant, and extremely sentimental bourgeois woman, as exemplified by Estela in *O cortiço*, Almeida
argues that the Estelas of this world could improve if they were
given the chance to receive a proper education and if they were
given serious, morally uplifting materials to read instead of silly
French novels.

As in most of Almeida's novels, *A casa verde* deals with a family,
a house, and a process of finding a suitable partner. Jeffrey Needell has made the clear connection between neocolonialism and
this idealization of French and British cultures.[76] Almeida's novel
recounts the story of the Lane family: Mr. Lane, an English businessman and widower, his daughter Mary, the French governess Mademoiselle Girard, and the Brazilian nurse Rita. Through
the protagonist and heroine, Mary Lane, a mix of tropical Bahia
and "Englishness," Almeida describes a utopian project in which
national identity could be harmonized with national progress.

The title of the novel is also the name of a house, *Casa Verde*, a
dilapidated structure located in Niterói, back then still a small rural
community north of Rio de Janeiro, that Mr. Lane bought, reno-
vated, and where he subsequently decided to live. By naming the
novel after the house, the domestic character of the work is highlighted; for Almeida, house and nation are integrally connected.
According to Mr. Lane, who calls the house Green House (in English)
despite his excellent command of Portuguese, family life is described
as far more important than his involvement in public society: "A
sociedade estraga os homens e avilta as mulheres. Só em casa, no

aconchego da familia, nos gozos simples, em que entram mais os exercícios materiais do que as preocupações fúteis que a vaidade sugere, se ampliam e aprofundam sentimentos nobres e que dão à consciência grande soma de benefícios"[77] (Society corrupts men and turns women into villains. Only at home, in the comfort of family, in the simple pleasures, in which material tasks matter more than the futile worries of vanity, do noble feelings grow and deepen and give the conscience a great many benefits). If we look at the domestic house as a laboratory for possible change,[78] it becomes apparent that Almeida is commenting on Brazil in a period when rapid economic and social changes led to feelings of insecurity. By focusing on the moral rectitude of Mr. Lane, she often suggests England is an example worthy to be imitated; however, as I will show, it is only through his biracial daughter that this potential can be fully realized.

In her study on the British in Brazil, Louise Guenther mentions that initially only male merchants and traders, who were said to lead "highly irregular" personal lives, were present. Later on, however, they were allowed to bring spouses and establish Anglican churches, thus initiating the development of a recognizably British community. The community's child-rearing practices were similar to those of British expatriate communities elsewhere in that those who could afford to send their children to boarding schools in Britain often did so. Nevertheless, several of these children returned as adults to live in Bahia. Daughters, in particular, may have preferred to live in Bahia rather than return permanently to Britain since young English ladies often enjoyed more social prestige within the expatriate community, compared to what they might expect in their home country.[79]

Physician Robert Dundas believed that the British body could not bear a stay of more than five to seven years in Bahia and noted that this problem was especially acute in females. According to Dundas, British women became physically ill in Bahia at a higher rate than British men. He found this noticeable enough to treat it in a separate section of his book, titled *European Females in Tropical Climates*, in which he claims the following: "This unexpected result must, I apprehend, be accounted for by the more indolent habits and mode of life of the former, favoured, if not altogether induced, by the languour inseparable from high temperature, sanctioned by the prevailing customs in most tropical climates, where household occupations are not attended to as in Europe, where fashion or custom precludes the enjoyment of active exercise abroad, and where even mental exertion is to some extent laborious."[80] If Almeida did not actually read this book, she was definitely familiar with such ideas, which may have caused her to regard Great Britain, more than France, as a healthy example for Brazil to follow.[81] The lives of actual British women in Brazil seem to have been seriously idealized by Almeida; according to Guenther, however, in many ways British women had fewer liberties than elite Brazilian women. While the latter were required to stay at home almost all the time, they were at least able to wear light, loose-fitting clothing. This the British women could not do. Guenther claims that Maria Graham's comments on the indecency of Brazilian women's house clothes illustrate this restriction.[82] Wearing tight clothes all day long and being unable to pursue the physical, mental, and social activities that presumably would have kept them at least as healthy as the men, it is no wonder that British women, unlike Mary Lane, frequently fell ill. In essence, they were paying the price—with their own bodies—of keeping up the behavioral demands of both British and Brazilian society.

In *A casa verde*, Mary Lane is without a doubt the heroine, but despite her physical beauty, she is not as attractive as might be expected. Mary is, in a sense, created by her father, without any female input:

> [Mr. Lane] não era homem de sentimentalidades e procurara por isso mesmo dar à filha uma educação libérrima, fazendo-a praticar a ginastica, natação, exercicios a pé e a cavalo a par dos estudos de musica, de desenho, historia natural e de linguas, que eram a seu ver, esteios magnificos para lhe ampararem a imaginação latina.[83]

> [(Mr. Lane) was not a sentimental man and for this very reason he sought to provide his daughter with an unconstrained education, making her practice gymnastics, swimming, horseback riding in step with her studies of music, drawing, natural history and languages, which were in his opinion, wonderful help in protecting her Latin imagination.]

However, for uneducated Brazilians, Mary often seems unsympathetic, even rude: "A originalidade do seu tipo e o seu modo aparentemente frio e concentrado faziam-na parecer antipatica às pessoas vulgares . . . Como o seu tipo, a sua alma era modelada pela influencia de duas raças fundidas. Impeto e reflexão; obstinação e piedade; independencia e meiguice"[84] (The singularity of her type and her apparently cold and concentrated manner made her seem rude to common folk . . . As her type, her soul was molded by the influence of two fused races. Impetuosity and reflection; obstinateness and piety; independence and tenderness). The education her father gives her, inspired by hygienic beliefs, is apparently in conflict with this Latin imagination, which for the widower is a mere "rede vulgar de preconceitos, superstições, fanatismos e vaidades"[85] (vulgar web of prejudice, superstition, fanaticism, and vanity). Lane wants his daughter to be strong and independent, not vulgar, vain, and superstitious. That is why he forces her to practice sports, in order to "fortificar o corpo, torna-lo ágil, independente, activo e belo . . . Assim como o seu corpo, ele sabia, ou julgava saber, que era o espirito da filha um espírito forte e perfeito"[86] (fortify the body, make it agile, independent, active and beautiful . . . Like her body, he knew, or thought he knew, her spirit was strong and perfect).

However, Mary is not exclusively British but *mestiça* (mixed race). She is Anglo-Saxon and Latina at the same time. Caught in between two cultures, she is subjected to constant misreadings by others, her father as well as the community in Niterói, but seems to be a very clear reader of others. Her Brazilian heritage, although barely mentioned through the figure of Mr. Lane's anonymous wife, will show itself in other ways and eventually enable her to better comprehend her surroundings, more so than her father ever could.

In *Desire and Domestic Fiction*, Nancy Armstrong suggests that the mother's surveillance within the family exerts a form of social control. To reframe this in Foucauldian terms, the mother plays the role of panopticon within the family. Thus, the mother's imposition of convention and quietude within the narrative opposes the need of the narrative for deviance and instability. Mary's mother is dead and, indeed, is not even named. Without her mother's absence,

Mary would have had to face conflicts of loyalty between her parents and may have been forced to make a choice between the two. Motherless, she will have to learn how to interpret her *mestiça* reality and clarify the multiple meanings of her Green House before she will be able to start her own family. In other words, had Mary had a strong mother to protect her from the villainous Guilherme Boston, or to dissuade her from her erroneous decision to keep up with the façade of her father's vision of Brazil, there would have been no story to tell.

It is made clear from the beginning that the Green House means different things to different people. The inhabitants of Niterói remember it as the scene of a brutal murder of an innocent woman hung by her insane husband; for a group of gypsies, it means easy access to financial gain; for Mr. Lane, it is a charming country home. Thus, the house itself becomes a space between cultures and cannot simply be seen as the civilizing element in the Brazilian landscape, as Mr. Lane believes. The Green House is the point where all paths cross, and it is the main goal of the villain, who wants to enter first to steal material objects and later to kidnap Mary, whom he desires to marry. Thus, the house, interchangeably called both Green House and Casa Verde, is not just the mark of British immigrants, a source of civilization and culture in a largely "backward" society in which, like a palimpsest, alternate realities compete for prominence. It is clear from the start that Mary is the only one who is able to read all the different meanings of the house. She happily exchanges her grandparents' home in Salvador for her father's in Niterói and is the only character who understands the elegant and tasteful renovations. However, she also hears the comments of her rather frivolous French governess, who prefers to see the "romantic" aspect of the dwelling, as she hears Rita, her maid from Bahia, voice her distaste for anything that is not from Bahia. Mary Lane is the one who invites the nun Pompéia, who helps her in keeping the presence of the gypsy body a secret, and eventually Guilherme Boston, who was using the group of gypsies to try to get inside. Mary discovers that her father's reading of the house as a safe haven is naïve and incorrect when she accidentally shoots a gypsy boy who was hiding in a large fig tree in their garden waiting for an opportunity to break into the house. The sexual symbolism is evident: she innocently shoots with a bow into a fig tree to kill an owl that was bothering her father, and all of a sudden a boy comes falling out. Mary decides she will keep him in her "quarto de virgem" (virgin's room) and thus initiates her own reading of this hybrid space.

Her father's sense of security is not to be violated, Mary decides, as she becomes increasingly aware of the conservative inhabitants of Niterói and their gossip about the Lane's Protestantism. Her own father mistakes her anxiety for homesickness and wants to send her to Rio de Janeiro. Only the handsome Doctor Eduardo Abrantes, who requires access to her room in order to take care of the injured Luis Ulka, quickly learns to "read" Mary correctly and helps her solve the mystery of the gypsy boy and eventually to find her true sense of self.

The union between the *mestiça* and the doctor is an obvious one, yet the father, unable to understand the local context, has set his mind on the villain of the novel, the Irish-Brazilian Guilherme Boston, as his son-in-law. Mary, a much more astute observer, immediately feels that she cannot trust him, and the reader soon discovers, through the dramatic tale of the innocent Laurinda, that under his apparent domesticity, Boston is a wicked villain who seduces and abandons innocent and defenseless girls.

The traditional equation between the body and the social corpus was linked to social Darwinism, eugenics, sexology, and all parts of the metaphoric discourse in which the physical body symbolized the social body. In this sense, physical and social disorder stood for social discord and danger. This ascribes a certain virility to doctors, who will be suitable spouses because they understand the female body. Thus, the physician's scrutiny and the lover's gaze become one. With such high stakes for masculinity, the profession is transformed into a spousal covenant when the physician becomes an expert on love. Doctors are established as the norm against which any deviance is measured. The final marriage between Mary and the doctor is celebrated as the fusion and mutual renewal of two different cultures.

CHAPTER 2

PURIFYING THE URBAN LANDSCAPE

PROCESSES OF IMMIGRATION, ACCULTURATION, AND RESISTANCE IN BUENOS AIRES

ARGENTINA'S "WHITENING"

One of the main ways in which the concern for Argentina's racial makeup manifested itself was the call for immigrants—specifically Northern European immigrants—by mid-nineteenth-century intellectuals. Their desire was to populate and make productive Argentina's vast lands in the interior. The hope was that any remaining indigenous peoples would be absorbed into the national body, and mestizos and criollos would be counterbalanced by "superior" Northern European stock. After the United States, Argentina had received the largest number of the European immigrants in times of mass immigration.[1] Between 1880 and 1890, more than one million immigrants arrived in Argentina; between 1890 and 1900, eight hundred thousand; and between 1900 and 1905, 1.2 million more people immigrated. Thus, by 1914, 30 percent of the national population consisted of immigrants who had radically transformed the culture and politics of Argentina and particularly that of Buenos Aires.[2] Arnd Schneider suggests that it is this particular era, from approximately 1880 to 1930, that we can consider Argentina's period of modernization, roughly coinciding with European modernity.

Settlement of the vast Argentinean pampas was a long-standing concern for national politicians, and the notion "to govern is to populate," originally coined by Alberdi, was a powerful trope that had long helped promote an open immigration policy. But by the early twentieth century, however, these policies were viewed much more negatively. As attitudes toward immigration began to change, "racial whitening" was no longer seen as an adequate answer to the problems of nationhood and identity; indeed, some people thought immigration should be severely restricted.[3]

Argentina seemed the only country in Latin America that had realized the ruling class's dream of transformation by racial whitening and cultural Europeanization. But racial ideology is not a matter of "racial facts" or social reality projected onto the realm of ideas. It is precisely in this transition period between the nineteenth-century idea of state and another more "modern" society that the concept of "inferior races" starts being used more frequently. If the debates about immigration are derived from the liberal Enlightenment notion of nation as a community imagined on the basis of common values, institutions, and political convictions rather than common descent, then at the turn of the century the debates about immigration come to focus on the process and the practices of assimilation or integration. According to William Acree, the question of whether to base the idea of the Argentine nation on citizenship and civic responsibilities or on ethnicity and tradition has been at the center of debate of Argentine identity and the nature of the Argentine polity at least since Bernardino Rivadavia. Fellow liberals attempted, unsuccessfully, to take the reins of the new republic of the United Provinces of the Río de la Plata in the early 1820s.[4] Supposedly, anyone (an immigrant from anywhere) could become a citizen, as long as he or she accepts the common values and agrees to live by the standards and basic codes of a particular national community. But this notion of assimilation is problematic because it assumes the resolution of difference: it promotes the idea of a stable, essentialist, unified national identity that supposedly absorbs, refines, and neutralizes difference but remains itself unchanged by those differences. Yet, in the cultural imaginary, it may be impossible to fully resolve all differences. Since some immigrants and their descendents might cultivate an identity distinct from the cohesively imagined national identity, there may always be some mark of difference that can be defined as the marks of a visual—that is, physical—difference, pointing to an essential, unassimilable Otherness. Under this mind-set, racial differences are ultimately irreconcilable.

Etiénne Balibar and Immanuel Wallerstein have argued that nationalist discourses can quickly result in violence when the "foreigner" within is forced out.[5] This violent expulsion is the outcome of what they call nationalism's inevitable relationship with racism, a relationship grounded on a fictitious concept of ethnicity. Racism is then not an expression of nationalism, but a supplement of nationalism or, more precisely, an internal supplement to nationalism— in excess of it, but always indispensable to its constitution. When people recognize each other as "members" of the same nation, such recognition inevitably aligns itself with the notion of a "race." Ultimately, Balibar and Wallerstein argue, the nation must align itself, spiritually as well as physically or carnally, with the "race," the "patrimony" to be defended from all degradation.

The emergence of a "Hispanic" conservative nationalism must be seen, I suggest, as a response to working-class cultures in factories, urban slums, and tenement buildings, known as *conventillos* in Argentina. It is important to remember that though Argentina was very close to becoming a truly "modern" country, a fear of being categorized according to an inclusive international category of "backwardness" always remained. It is this fear of Buenos Aires not being quite civilized, read "national" enough, that links the texts analyzed in this chapter. I will open with an analysis of a sociological text, *Las multitudes argentinas* (1899) by José María Ramos Mejía, and conclude by studying two novels: *La bolsa* (1891) by Julián Martel and *Stella*: *Novela de costumbres argentinas* (1884) by the all-but-forgotten Emma de la Barra. All these texts continue the debate on the benefits and dangers of immigration.

To explain Argentina's debates on racial thought, I start with a brief overview of the changes in the construction of the nation and the way these changes informed the politics of national identity. As Halperín Donghi points out, the assimilation that Sarmiento and Alberdi expected from immigrant groups will constitute one of the ideological cornerstones of positivist thought until the Argentine

Centennial in 1910.[6] Immigration was no longer a future program but a policy that had changed Argentine society beyond recognition. The immigrant was no longer received in an empty space, but in a national typology that was being destroyed. It is precisely the fact that foreign, unknown elements began to occupy a previously familiar landscape that proves so threatening to these foundational figures. Sarmiento himself, once one of the main promoters of the whitening immigration policy, shifts his attitudes drastically after the 1880s and admits that his hope in immigration as an instrument for progress had not been as successful as he assumed in the days of writing *Facundo*. A fear of social and labor activism contributed to the decreasing "desirability" of immigrant groups altogether, eventually overshadowing issues of "racial improvement." Thus, the immigrants were connected to two threats in the Argentina newly transformed by modernity: the laboring masses and the rising middle class.

As Beatriz Sarlo points out, the city, initially seen as the site for progress, was now increasingly thought of as a symbolic and material condensation of these social changes.[7] Finally, the city had conquered the countryside, and immigration caused significant demographic changes. Within a single lifespan, the small city of Buenos Aires had been transformed into an industrial conurbation. The impact on human sensibility was all the greater in the case of Buenos Aires, Sarlo explains, because so many city-dwellers were not born in the metropolis but were immigrants from more traditional communities.[8] These rapid changes deeply influenced the way the past was constructed because the population could remember the city when it was quite different. This was the city of their youth or adolescence, and the biographical past underlines what has been lost or changed forever. A city that doubles in size in less than a quarter of a century undergoes changes that its inhabitants, old and new, have to process, while the countryside, in turn, is felt like a wound, a reminder of all that has disappeared.

Many of the immigrants, it had turned out, were not of desired European ancestry but were of "eastern" or "Oriental" origin (meaning Syrian, Lebanese, or Russian—often meaning Jewish) whose capacity to assimilate into the Argentine way of life was profoundly doubted by the old elite. The failure of many working-class immigrants to become Argentine citizens or to marry outside their own immigrant circles added to the sense among the elite that foreign "cysts" were forming in their midst, endangering the national body. Meanwhile, the new middle class also challenged the traditional political system of representation. To many conservatives, the Argentine culture was giving way to a threatening mix of peoples. Such fears generated xenophobia and inflamed a desire to defend true "argentinidad" ("Argentineness" or Argentine identity) uncontaminated by alien elements.

Labor unionizing and class struggle itself was seen as "artificial" and "exotic," and for the elites, labor protests were increasingly seen as a consequence of these new immigrant cultures mixed with immigrant issues. The Argentine population was now seen as divided into "exotic" masses, linguistically and culturally cosmopolitan, as opposed to "native" criollos, speaking Spanish.[9] Though one government strategy was to try to assimilate these immigrants—at least as concerns culture and ideology—as quickly as possible, the combination of their political radicalism, most often described by the authorities as "foreign agitation," their upward mobility, and their range of ethnicities led to a serious backlash of conservatism with racist overtones. This backlash would eventually take the form of a nationalism defined in large part by its "rediscovery" of the country's foundational, Hispanic, criollo, or gaucho "origins" and its antiliberal hostility toward urban immigrants and internationalism in general. This nationalist ideology, proposing a direct correspondence between individual, language, and nation, defines citizenship in exclusionary terms. Purity of language and morals were sought with equal fervor as "purity" of civic allegiance and patriotism.

Lilia Ana Bertoni notes that in Argentina the definition of citizenship itself, together with the naturalization of immigrants, was intensely debated.[10] There was uncertainty among policy makers and immigrants alike about what exactly naturalization and Argentine citizenship should entail. Naturalization itself was clearly seen as a necessity for the immigrant's integration into the nation, but what citizenship required was less clear.

In 1902 the Residency Law was approved, authorizing the expulsion of foreigners who threatened national security or public order that allowed the extradition of—or refusal of entry to—any foreigners who could be associated with the vaguely construed category of "common crimes." The law, whose goal was to halt social agitation, activism, and worker violence, only made specific reference to foreigners. All foreigners who participated in union activities could be deported. The Law of Social Defense of 1910 continued this identification of immigrants as labor activists and explicitly refers to all terrorists as foreigners. According to Juan Suriano, this law was based on "pseudo-scientific" studies that considered anarchist thought not as a social phenomenon but rather as a perverse, morbid, and deviant nervous affliction of individuals who were expelled from their own countries, where these ideas abounded because misery and alcoholism had affected the population's brains.[11] The Sáenz Peña Law, sanctioned in 1912, ensured political participation through mandatory vote but did not help the democratic process since the right to vote was limited to native-born males only. Women and immigrants were thus formally identified as the nation's "Others." If immigrants desired to obtain Argentine citizenship, these laws made it extremely difficult for them to do so.

By 1914, more and more foreigners were becoming politically active, and members of the ruling elite increasingly tried to justify the exclusion of immigrants from politics by pointing to their cultural inferiority. Calls to streamline the complex naturalization process were ignored. Even naturalization was not always a guarantee of equal treatment between foreign-born and Argentine-born nationals, especially in judiciary and diplomatic careers.[12] By accepting these restrictive laws, the ruling classes thus showed the limits of their liberal ideology and the equation of social unrest with "foreign elements." To distinguish the immigrants from their supposedly more illustrious ancestors, the ruling class differentiated between "Gallegos," penniless contemporary immigrants, and "conquistadores," the "valiant founders" of Argentine culture. There were not as many blacks or mulattos in Argentina compared to Brazil; therefore, according to Julio Ramos, by the turn of the century, the dominant elites were beginning to mark southern European immigrants with the stereotypical metaphor of "blackness."[13]

Initially, immigration was oriented toward agriculture and was supposed to help develop the productive potential of the pampas.

But the distribution of land in Argentina had always been very limited, with the main colonies outside of Buenos Aires established near Córdoba and south of Santa Fe. The vast majority of new-comers were forced to seek employment in urban areas. As Gladys Onega points out, this nascent proletariat, which could not be absorbed economically or socially, soon began to participate in union movements and labor strikes, adhering to the tenets of both socialism and anarchism, participating in attempts at armed revo-lution just after the turn of the century, and helping to bring the populist, reform-minded Radical Party to power by 1916.[14]

It is important to remember, as Halperín Donghi reminds us, that all these xenophobic measures that justified the repression of labor movements and social protests never resulted in a signifi-cant modification in the immigration policy itself.[15] It is precisely around this period that the number of immigrants reached its peak. Instead of serving as a blind instrument in the hands of the elite, immigrants started to improve economically, competing in the labor market, forming national organizations, and dealing with exclusionist notions of Argentine identity.

PASSING FOR WHITE: *LA BOLSA*

La bolsa by Julián Martel (pseudonym of José María Miró) might be best remembered for its flagrant anti-Semitism, but I would suggest that the immigration politics portrayed in the novel are more complex than generally presumed, and rather than being exclusively anti-immigrant, the novel questions the possibility of a homogenous and singular nation. Because of the elimination of the indigenous peoples and the small number of Afro-Argentines, racial differences such as skin color between "natives" and "immi-grants" were hard to distinguish. It seemed that the foreigner's best defense from discrimination was to hide behind a mask of national respectability, simulating a civilized and civilizing patriotism while hiding his or her inner monstrosity.

Immigrants were, in this view, a kind of virus that infected the body of the nation. In such a pathological rhetoric, the anthro-pologist figures as a medical researcher, tirelessly collecting and classifying data among the cells of the social organism. Mean-while, anthropological discourse gains legitimacy by providing the institutions of power with a therapeutic and prophylactic gaze on the dangerous classes. In that sense, as Francine Masiello observes, the virus operates as a self-constituting other, as the condition that paradoxically enables the constitution of the healthy national subject.[16] The therapeutic gaze of the anthropologist responded to such a crisis by representing and ordering the flow of people. In many ways, the array of criminal traits was the "scientific" effect of a sinister hermeneutics that interpreted the physiological traits of ethnic difference as signs of social and moral disorders. In responding to the question of how "we" (the "healthy" citizens of the nation) can recognize a dangerous "other," the anthropologist-detective morbidly collects all kinds of physical data.[17]

How can the good citizen possibly be differentiated from and protected against hidden degenerating elements? That seems to be the main question in *La Bolsa*. The novel is set up as a set of tableaux showing the downfall of an honest and good man, the son of impoverished English immigrants, Dr. Luis Glow. Like Jerônimo in *O cortiço*, he too is a sort of larval character, a potential holder of civilization, hence his name, Glow, which refers to light that can be seen as the light of civilization. Through honest hard work, Glow dramatically improves his social status. Upon graduating with a law degree, he establishes a large clientele, marries a virtuous wife, and obtains a luxurious house. Unfortunately, he is seduced by the prospect of easy money to be made at the stock exchange, *la bolsa*, and becomes involved with risky speculations. The naïve Glow trusts several unsavory characters, participates in questionable deals, and ends up losing absolutely everything, including his sanity. The threat to Argentina in the novel is not immigration per se but the increasing difficulty of distinguishing between the good and the bad immigrants. The presence of Glow himself and the good Italian stockbroker Carcaneli suffice to dismantle that argument. Right after the vicious description of the Jewish Baron Macksler, there is another description of Carcaneli, who, in a fate similar to Glow's, will die broke and insane in Italy:

> Aun hoy se ve, en el centro de la Avenida República, el palacio extravagante que edificó en el apogeo de su fama y de su fortuna, y que demostraba, por la rara disposición de su jardín estrambótico, muy cambiado ahora, el desorden mental que empezaba a trastornarlo,

acosado por la ambición frenética de llegar a ser el árbitro de las finanzas argentinas . . . Era grande en todo. Generoso, bueno, espléndido, amado de la juventud, a quien estimulaba y protegía.[18]

[Even today one can see in the middle of the Avenida República, the extravagant palace that he built at the peak of his fame and fortune that showed, through the strange disposition of his outlandish garden, nowadays much changed, the mental disorder which started to disturb him, hounded by the frenetic ambition to become the referee of the Argentine financial world . . . He was grand in everything. Generous, good, splendid, an admirer of the younger generation, which he stimulated and protected.]

Given this ambivalent attitude toward immigration, it seems that *La bolsa*'s much commented, nasty anti-Semitism is more a result of French literary and cultural influence than any concrete political program. I have found no specific anti-Semitic laws in place at the time, nor any systematic action taken against Jews by the government, and the outbreak of pogroms in Argentina did not occur until 1917, twenty-five years later. The big fear expressed in the novel is precisely the lack of individuality that results from modern life. As George Simmel notes, it is money, with its colorlessness and indifferent quality, that becomes the frightful leveler of previously distinct social classes.[19] The development of an exchange economy simultaneously encourages a form of rationality that reduces qualitative difference to quantitative difference and promotes a calculating intellect at the expense of emotional impulses. While in certain respects modernization promotes difference and diversity, it also, paradoxically, engenders an increasing sense of the homogeneity and interchangeability of objects and persons. Modernity, according to Simmel, brings with it an inexorable rationalization that lends to a weakening of sentimental attachments and personal ties, resulting in social fragmentation.

Martel violently rejects this loss of identity by juxtaposing it with a coherent (inherited) national tradition with a unified sense meaning. If everybody can become a "*Porteño*," a person from Buenos Aires, and money seems to be the only common language left, then what possible basis could there be for a unique national identity? How can one possibly look for and study the national essence if society changes so quickly? Paraphrasing Marshall Berman, one could say that the old modes of honor and dignity do not die; instead, they are incorporated into the market, take on price tags, and gain new life as commodity.[20] Thus, any imaginable mode of human conduct becomes morally permissible the moment it becomes economically "valuable"; anything goes, as long as it pays. The standardization of values that the circulation of money demands implies that relations among individuals can be likened to an exchange of commodities and goods, described by Simmel as a cold intellectual process.

In the novel, public space is associated with a fear of contamination and disorder arising from a leveling of class distinctions: "Uno de esos edificios tan comunes en nuestros barrios centrales, construidos con el linceo propósito de sacar de la tierra el mayor beneficio posible, sin tener para nada en cuenta el gusto arquitectónico ni los preceptos higiénicos relacionados con la acción del aire y de la luz sobre el organismo humano"[21] (One of those often-seen buildings in our central neighborhoods, built with the single objective of gaining the greatest possible benefit from the land, without considering at all architectural taste or hygiene concerns related to the effects of airflow and light upon the human body). At the theater and at social events, the anonymity of the crowd and the promiscuity of individuals subvert established social divisions; hierarchies are undermined in the public domain as disparate individuals rub shoulders in the common pursuit of pleasure.

The stock exchange generates money, and where money abounds, patriotism is rare. The immigrant who comes for material gain will likely not be interested in the progress of Argentina as a nation. The danger, according to Martel, is not that the immigrant is different, but rather that he starts looking and behaving too much like the local-born man. Race will not reveal this dangerous alterity, nor will social class. Since *Porteño* life has hopelessly fragmented, social classes intertwine in a manner never seen before, and people are never what they seem to be.

Beatriz Sarlo asks, who exactly was an Argentine in that period?[22] She suggests the defining factor is neither race nor socioeconomic class, but something that is much harder to hide: the question of language, of who speaks and writes an "acceptable" language. It is here that the immigrants reveal their true selves: the accent that they can never fully disguise: "Fouchez hablaba el español con bastante claridad, aunque su pronunciación gutural, unida a cierta petulancia muy propia del carácter francés, denunciaban su origen"[23] (Fouchez spoke Spanish reasonably clearly, though his guttural pronunciation, together with a certain vanity characteristic of the French character, revealed his origin). Worse still is the case of the villain of the novel, Baron Mackser:

> El que hablaba masticando las palabras francesas con dientes alemanes, y no de los más puros, por cierto, era un hombre pálido, rubio, linfático, de mediana estatura, y en cuya cara antipática y afeminada se observaba esa expresión de hipócrita humildad que la costumbre de un largo servilismo ha hecho como el sello típico de la raza judía.[24]

> [He who spoke chewing French words with German teeth, and not of the most pure, by the way, was a pale man, blond, lymphatic, of medium build, and in whose disagreeable and effeminate face one could see an expression of hypocritical humility that a long history of servility had turned into the hallmark of the Jewish race.]

Mackser seems to embody the worst fears of Argentine nationalists of the period. He is a moneylender but strangely enough also a labor union organizer. Sexually, he is also deviant, with an effeminate face, and it is suggested he might have a homosexual relationship with "un joven, compatriota y correligionario suyo, que ejercía el comercio de mujeres, abasteciendo los serrallos porteños de todas las bellezas que proporcionan los mercados alemanes y orientales . . . Pálido, rubio, enclenque y de reducida estatura, sabe Dios qué extraños lazos lo unían con el barón de Mackser, al que parecía tratar con exagerados miramientos" [25] (a young, like-minded comrade, who supplied the *Porteño* brothels with all the female beauties that the German and Asian markets could offer . . . Pale, blond, sickly, and short, God only knows what strange ties connected him with Baron Mackser, whom he seemed to eye excessively). For Martel, then, not all immigrants are bad; Glow and Carcaneli were essentially good people who were seduced by money. Immigrants who work for the improvement of Argentina as well as for themselves are necessary in Martel's logic. However, his fear is that the bad ones will overrun the country through the power of their money, destroying all local traditions. Money and personal greed are everywhere in the city, and individuals such as Mackser increasingly gain control.

THE DYSFUNCTIONAL NATIONAL FAMILY

In her study of Argentina's naturalist novels, Gabriela Nouzeilles suggests that narratives of the 1880s reveal the first shadows of skepticism about the conciliating figures described by Doris Sommer in the Romantic novel.[26] These new Naturalist novels would also adopt the form of allegory, but not to show the obstacles that forbade the achievement of a national utopia. Rather, these novels question the possibility of such a unified national type. If the arguments of the foundational novels of the early nineteenth century were projections of the strategies necessary to reach a political consensus, the Naturalist arguments denounced the fragility of such agreements. Francine Masiello has noted that *La bolsa* opens with the pairing of natural and domestic violence, when a violent, torrential downpour is likened to a bilious husband intent on beating his wife.[27] This scene leads the reader to expect further disturbances in the Argentine family. In other words, the natural state of the narrative will be found in scenes of domestic trouble. *La bolsa*, however, does much more than survey the Argentine family; it studies the family as a nation model for its deviance and ability to disrupt.

Even Glow's virtuous wife, one of those "almost extinct Andalusian beauties" and a daughter of heroes of the country's independence wars, is unknowingly corrupted by her need for luxury. The description of Glow's house, with its Japanese plants and mosaics worthy of an Oriental palace, should thus be read as a premonition of things to come, especially since the stock market itself is also described in feminine, Orientalist terms: "la grande, generosa, opulenta, adorable Bolsa, dispensadora de todos los beneficios, cueva de Alí-Babá y lámpara de Aladino"[28] (the great, generous, opulent, adorable Exchange, dispenser of all goodness, Ali Baba's cave and Aladdin's lamp).

Female desire for consumer goods is incited principally by the production of visual pleasure. Because of their reproductive qualities and supposed stability, women cannot truly belong to modernity in the same way as men. By dressing herself in luxury, however, she can become a seductive creature of artifice. On the final page of the novel, Glow, who upon losing his fortune has also lost his sanity, sees his wife's face turn into a mask and then disappear. It is replaced by a seductive Cleopatra who abruptly transforms into a horrible Medusa: "Y él entonces, debatiéndose en el horror de una agonía espantosa ¡loco, loco para siempre! Oyó estas tres palabras que salían roncamente por la boca del monstruo: soy la Bolsa"[29] (And then, debating in his mind the horror of a terrifying agony, he went crazy, crazy forever! He heard these three words that came roughly from the monster's mouth: I am the Stock Exchange).

Woman *is* the stock exchange, the devouring monster, the embodiment of modernity. The consumer too is portrayed as a woman (in this case, Glow's wife), and feminization comes to equal the demonization of modernity. Women are portrayed as buying machines driven by nothing more than their impulses to squander money. Economic and erotic excess become closely related, and men are feminized by the castrating effects of an ever more pervasive commodification of Argentine society. No longer equated with a progressive development toward rationality, man is seduced by marketing techniques and is no longer in control of his desires. The masculine robust sense of individual self has been invaded and feminized by this omnipresent modern culture.[30] Modernity now comes to exemplify the growth of irrationalism and the return of a repressed nature in the form of inchoate desire. The metropolis comes to be depicted as a woman, a demonic femme fatale whose seductive cruelty exemplifies, simultaneously, the delights and horrors of urban life.

THE CHANGING COUNTRYSIDE

The escape from modernism that *La bolsa* suggests can be found in the countryside, as of yet untouched by modernity. The quest for better models, new forms, fresh images, and relief from the ills of metropolitan centers compelled modern man to move to areas then perceived to be at the margins of the "civilized" world. The countryside suddenly became an ideal space outside of modernity, so that it eventually came to mean "true" civilization rather than barbarism. Barbarism was soon identified with the city, in the form of the European immigrant, who hailed from Europe's barbarous rural zones and who had come to take over the American city.

The desire to return to the countryside can be attributed to the fragmentation of the modern age, as increased mobility and demographic shifts caused significant sectors of the population to be uprooted from their native lands, and hence lose a sense of continuity with their birthplace and history.

Rather than populate the interior, as Alberdi and Sarmiento had envisioned, most immigrants stayed in Buenos Aires. The city, populated by numerous poor immigrants prone to labor-organizing and a growing middle class contributed to establishing a climate of fear about foreignness and materialism among the established classes, as well as a sense of nostalgia for the countryside, lost values, and pre-immigrant Argentina. In this sense, we can use the term "cultural nationalism" to describe the concern expressed by the ruling elite, which saw their society as threatened by disintegration of modern life and the specifically antimodern attitude espoused by the provincial land-owning class, which felt threatened by the rapid pace of modernization. Groups of intellectuals started to reject positivist and cosmopolitan ideas in favor of nationalistic ones that defended traditional Hispanic culture and eschewed the "cosmopolitan flood" and the "Babel of languages" that in their view characterized early twentieth-century Argentina. These intellectuals cast themselves as heirs of the prerepublican "national essence," and many of those who had ardently embraced European culture as a model now began to nostalgically look to their criollo past, preferring Spanish customs and what used to be called the barbarity of the countryside. This ideological shift led to a clear break for the republican, immigrant-sympathizing founding fathers of Argentina. Rural areas, against which the republic had previously based its self-definition, were now embraced ardently by nationalist intellectuals and artists who recuperated this space to the service of a triumphant ideology.

These ideas will gain momentum during Argentina's centennial celebrations in 1910, but already, in Martel's novel, the countryside is equated with a truly "Spanish" tradition from which the immigrant is excluded. The author only briefly mentions Granulillo's brother, the innocent farmer Lorenzo, who sells all the land he owns in order to travel to Buenos Aires and try his luck investing in the stock market. Once in the city, unsurprisingly, he loses everything to his own brother.[31]

LAS MULTITUDES ARGENTINAS
BY JOSÉ MARÍA RAMOS MEJÍA.

In its paranoid fear of modernity, *La Bolsa* directly predates the thoughts of José María Ramos Mejía (1849–1914), a scientist, doctor, educator, writer, and pioneer of Argentine psychiatry. Perhaps his most famous work, *Las multitudes argentinas* was published in 1899 and makes for a fascinating read with its bizarre mixture of racism, fear, and condescension.

Hugo Vezzetti comments about this period:

Es posible rastrear extensamente en las producciones científicas y literarias argentinas, del 80 al Centenario, . . . un darwinismo vulgarizado . . . si el progreso social aparece fundado en las leyes irreductibles de la biología, no deja de pagar un precio: la figura siniestra de la degeneración condensa el negativo exacto de los valores morales deseados. Un lazo sólidamente significativo liga esa noción, asociada a la locura urbana, por una parte a una reedición de la barbarie en la gran ciudad, y a la vez a una expresión agudizada de los conflictos de la inmigración.[32]

[It is possible to find in the scientific and literary output of the 1880s to the Centenary, . . . a vulgarized Darwinism . . . if social progress is anchored in irreducible laws of biology, there is a cost: the sinister figure of degeneration condenses the exact negative of the desired moral values. There is a strong link between this notion (of degeneration) to urban insanity; one the one hand because of a rewriting of barbarism in the city and at the same time heightened confrontation of the conflicts surrounding immigrants.]

For Ramos Mejía, positivist ideology provided not only an interpretation of national reality but also a tool for understanding educational, sanitary, and military questions. Ramos Mejía attempts to explain the unforeseen and unwanted effects of immigration in Argentina as he looks for remedies. In his view, the relation between the state and the masses will define the future of the nation, thus clearly repositioning the question of the so-called National Problem. In this aspect, he follows thinkers such as Gustave Le Bon, who investigated the "popular imagination" and declared that one had to know the way the imaginary of the masses worked in able to better control and govern them.[33] The crowd comes out of the masses, and the mob comes out of the crowds—all at the expense of the community. Not just an aggregate of individuals, the crowd has a life of its own: it becomes a sort of atavistic animal and is capable of bestial and violent behavior, mindlessly following a leader.

Nevertheless, unlike Le Bon, Ramos Mejía does not believe that each and every man can form part of this mass. Unlike in France, of course, the typical member of the lower-class public in Argentina is an immigrant, of no great intelligence, who thinks with his heart or stomach, not with his head. In the first chapter, Ramos Mejía studies the role of the masses in the Argentine political climate. He conceives of these "masses" as if they were a specific type of cyst, inserted into the country's body.[34] Although it may be more agreeable to analyze the deeds of great men, for Ramos Mejía, it is precisely the vague and diffuse face of the masses that needs to be examined. Only then can we decide what doses of which medicine to administer to the country.

In a group, the feelings and ideas of each individual become diluted and transformed into what the author calls "el alma de la multitude," or the soul of the masses. This multitude forms one single subject, albeit provisional, which is constituted by heterogeneous elements that are but momentarily unified. For Ramos Mejía, the masses are cells, starting to form a living body and a new and different being. This process of "massification" should be seen as a psychological adaptation, because certain characteristics of the contaminated individual will be deformed or destroyed, whereas others will become greatly exalted. One should see the man of the masses as a creature of pure instinct: impulsive, aggressive, and animalistic, although he can also be heroic and generous at times. But intelligence, reason, or other rational qualities will be totally absent in him, as these can only be found in a reflexive man.[35]

If the modern man in Europe, so cultured in isolation, can become such a barbarian when he forms part of the multitude, it goes without saying how barbarous Argentine masses, already more instinctive and primitive, can become: "¡Qué no serían estas nuestras informes colectividades, sin el secreto freno de la fuerza de inercia que da la civilización inconscientemente en el cerebro!"[36] (What would our informed collectivities be without the secret restraint of the force of inertia that the sum of our civilization unconsciously gives in its mind!).

The masses have no internal coherence except for this momentary connection that transforms different individuals into an impulsive community, causing them to perform heroic as well as criminal acts: "Constituyen los principales núcleos de la multitud: los sensitivos, los neuróticos, los individuos cuyos nervios sólo necesitan que la sensación les roce apenas la superficie, para vibrar en un prolongado gemido de dolor en la vigorosa impulsividad, que es la característica de todas las muchedumbres"[37] (They constitute the principal nuclei of the multitude: the sensitive, the neurotic, those individuals whose nerves need only to be grazed on the surface, in order to shake in a prolonged cry of agony in their vigorous impulsivity, which is the [predominant] characteristic of all vulgar masses). For Ramos Mejía, the masses are characterized as feminine, and he compares these "massified" creatures to amorous women, pulsing with instinct and passion, full of light, lovers of violent sensations, bright colors, loud music, handsome men, and big statues, sensual and perverse. He is hardly alone in this perception of modernity as female: shaken by this vast immigration from Europe. By the beginnings of industrialization, scientific research, and modernity, Argentine intellectuals increasingly portrayed women as responsible for the intrusive evils of prostitution, profits, and money. As an example of the degradation of an idealized European model, the female was invoked to symbolize the failure of the Argentine program of enhancing (that is, whitening) the race. Ramos Mejía literally compares the foreign masses to a woman in love: irrational, irresponsible, hysterical, and in need of close surveillance.

Fortunately, just like this lovesick woman, the masses are not very intelligent, cannot reason, and occasionally transform into

un perpetuo gongorismo moral, ampliando y magnificándolo todo
en proporciones megalomaniacas. Enamorada de la leyenda de cuyo
color vive, toda se convierte entre sus manos en cuento de hadas o
en fantasías . . . porque careciendo de contrapeso de las funciones
superiores del espíritu, todo lo entrega a la sensación y a la tenden-
cia de supersticiosa grandeza.[38]

[a perpetual moral *Gongorism*, amplifying and magnifying everything in megalomaniac proportions. In love with legends, from
whose color she lives, everything, in her hands, becomes a fairytale
or a fantasy . . . for, lacking the counterweight of superior spiritual
faculties, she turns all over to sentiments and to the tendency of
superstitious greatness.]

For positivists, women had always lacked the physical strength
with which to compete in the struggle for life; therefore, as Karen
Mead has shown, their evolutionary survival had depended on the
refinement of fraudulent means to defend against the brute force
of men.[39] As Mead explains, improperly channeled sexual passion
also constituted a threat to contemporary society in the cities.
The immigration of European laborers and the dramatic increase
in prosperity had not automatically produced the desired social
developments. Modernity also inspired fears of degeneration, particularly dreaded in Buenos Aires, where the sexual imbalance and
unhealthy materialism linked to immigration were most evident.
Interestingly enough, Mead shows how the disorder of the urban
underclass was increasingly associated with female hysteria.[40] Thus,
a predominantly male immigrant population becomes increasingly
associated with a hysterical woman.

In order to form a mob, it is necessary to have a certain psychic
structure, as a type of contiguous contact is indispensable, which
means that isolated individuals can transform into an explosive and
threatening phenomenon. Ramos Mejía dedicates several chapters
to the development of the Argentine masses in the colonial period
and the early nineteenth century. These "native" Argentines maintained their Spanish character and show a tendency to support tyr-
anny, eventually founding the nation's coastal cities. This man from
the coast

sale de los litorales del Río de la Plata, y es indio, heterogéneo como ninguno y completamente inculto, es casi autóctono mestizo-español en parte, y constituida por el hombre de la naturaleza que se ha formado en la soledad y el aislamiento de los desiertos inmensos y en los montes sin fin.[41]

[comes from the banks of the Río de la Plata, and is an Indian, heterogeneous like no other and completely uneducated, he is almost autochthonous mestizo—Spanish in part, and constituted by the man of nature who has grown up in the solitude and isolation of the immense wilderness and endless brush.]

The people from the interior, in contrast, are described as purely physical, with very little intellect, more tolerant, and slow to assimilate but very religious if not outright superstitious:

La verdad es, que cuando de esta ciudad multicolor y cosmopolita en demasía, uno se traslada a la tranquila ciudad del interior, siente que el alma levanta sus alas suavemente acariciada por el recuerdo de la vieja cepa; percibe algo que semeja la fresca brisa de la infancia cantando en la memoria multitud de recuerdos amables.[42]

[The truth is, that when from this excessively cosmopolitan and multicolored city, one moves to the quiet city of the interior, one feels that his soul opens its wings delicately caressed by the memory of the old stock: perceives something similar to the cool breeze of childhood singing in his memory a bevy of agreeable reminiscences.]

Ramos Mejía imagines Argentine history as two streams, one from the coast and one from the interior, both running toward the immense capital:

la Capital fenicia y heterogénea todavía, pero futuro crisol donde se funde el bronce, tal vez con demasiada precipitación, de la gran estatua del porvenir: la raza nueva. Por esto, aunque lentamente, va resultando cierta unidad de sentimiento político entre la metrópoli y el resto de la República; y precisamente por esa la multitud que se forme aquí tendrá más tarde su tinte nacional.[43]

[the Phoenecian and still heterogeneous capital, the future bronze-forging crucible, perhaps all too quickly, of the great statue of the future: the new race. This is why, though slowly, a certain unity of political thought between the metropolis and the rest of the republic is forming: and precisely because of this unity, the multitude that is formed here will later reveal its national color.]

The capital is the brain, the intellectual center of the nation and body. Here the new Argentine is born. This process is comparable to the evolution of a primitive fish into a human being, Ramos Mejía claims, and then immediately compares the immigrant to this primitive fish. Everybody is smarter than the recently arrived immigrant, whom he calls amorphous, cellular, resembling the oxen with which he has always lived. He recalls that he was once in the "Asilo de Inmigrantes" (Immigrant Asylum) where he did psychological experiments that allowed him to reach the following conclusions: "Me asombra la dócil plasticidad de ese italiano inmigrante. Llega amorfo y protoplasmático a estas playas y acepta con profética mansedumbre todas las formas que le imprime la necesidad y la legítima ambición"[44] (The docile plasticity of this Italian immigrant astonishes me. He arrives amorphous and protoplasmic to our beaches and accepts with prophetic tractability all the forms that need and legitimate ambition impress upon him). Since there are so many of these amorphous immigrants, they overwhelm the city, the second- and third-rate theaters, the churches (as they are very religious), the streets, the markets, and the hospitals. They are suitable for low-level jobs that do not require intellectual skills:

Porque, en efecto, ese desagradable . . . no es sino un símbolo vivo del inmigrante italiano . . . Así, le veis en ocasiones, marido fiel y constante de una paisana, amante de una negra o rendido amador de una china suculenta o de alguna solterona centenaria, cuyo capitalito sin movimiento él fecunda.[45]

[Because, in effect, this disagreeable (individual) . . . is nothing more than a living symbol of the Italian immigrant . . . Thus, you see him at times, a loyal and constant husband to a likewise Italian wife, lover to some black woman, or turned devotee of a succulent young girl or of some ancient old maid whose unmoving little capital he inseminates.]

In the countryside, Italian immigrants have even occasionally displaced the gaucho, riding a horse and working the land. Some go so far as to dress themselves up as gauchos, to the great astonishment of the locals who would never do such a thing themselves. According to Ramos Mejía, from the first generation there is nothing to fear; this larvae-like immigrant seems to be a very good and humble man. His child will inherit greater aptitudes than if his parents had stayed in their country of birth. Nevertheless, this biological heritage is too weak, and the *pilluelo*, or Argentinized child, will inherit local, Argentine characteristics. Restless and curious, always in the street, it is necessary to instruct this child in Argentina's heroic past, its national hymn, and reverence for the flag.[46]

According to Ramos Mejía's description, the children of immigrants, with their odd mixture of names and surnames, come from an unknown nest, and the author invokes both racial and sexual stereotypes in his characterizations. Descriptions of the first generation are reminiscent of black stereotyping: the children are often physically deformed and ugly, as if they were shaped by a rough mould, with flat noses, big ears, and thick lips. But the second generation already looks much more civilized, because of the better food and air of Argentina and a less agitated life that help them develop their intelligence more than their ancestors. Indeed, these *guarangos*, descendents of immigrants, can actually rise in social position. Still, Ramos Mejía compares this group with the *invertido*, or sexual deviant. Indeed, he thinks, *guarangos* and *invertidos* are remarkably similar, both love shrieking and dressing in vivid colors. Only *invertidos* and *guarangos* (and women, one might add) have this peculiar taste. The *guarango* can be a social climber with access to high society, as he attempts to hide his vulgarity and humble origins. Yet this figure disappears after the third generation and is followed by the *canalla* (literally, "pig") and the *huaso* (literally, "peasant"). These are often dishonest individuals and would normally be feared were it not for the benefits of a strong national education.[47] Thus, Argentine nationality should be created by carefully combining political, social, and economic influences mixed with the mold of this special environment. One should not interrupt this process with radical or brusque interventions.[48]

In Ramos Mejia's vision of Buenos Aires, there is no more hunger, no misery, no hate, no unemployment, and no families without heat in the winter. The epidemic of misery that entrenches a personality based on disease, in addition to dangerous social phenomena, such as those that can be seen in Europe, does not yet exist in Argentina. Admittedly, some alcoholics and vagabonds can be seen, but if the body is satisfied and the mind not worried by troubles, the sensual, hysterical, terrifying mob will not come. Buenos Aires still thinks too much with its stomach, like Sancho Panza, and the day that the masses go hungry, socialist groups will quickly organize and contaminate the city.[49]

The automatic, instinctive behavior of the immigrant can be modified through mental development, primarily through education. The Italian, the *guarango*, the *invertido*, and the immigrant dressed as a gaucho are all the opposite of Ramos Mejía himself. He is a real Argentine, whereas they belong nowhere and have surnames that are impossible to pronounce. In this manner, all Others are rendered homogenous, "barbarian," incomprehensible, and therefore, ultimately unimportant for Argentina's future.

For Ramos Mejía, masquerading as a gaucho is acceptable, even funny, if it is taken as a sign of admiration for the host country. It is only when this act is performed too well, when the immigrants become too sophisticated (and undetectable), that the process becomes threatening. Ramos Mejía, in conclusion, as both an anthropologist and a medical researcher, is a man who tirelessly collects and classifies data among the cells of the social organism. His notion of national character implies the existence of a hidden and essential characteristic that would distinguish between "us" and "them."

AN IMMIGRANT CINDERELLA:
STELLA: NOVELA DE COSTUMBRES ARGENTINAS

Obviously, in a nation concerned with maintaining racial purity through a line of legitimate heirs, the situation of unassimilated women presents a problem. In general, once a nation is founded and its land domesticated, the nation becomes the motherland. Metaphorizing geography in such a way posits a primal, essential, universal, and emotional linking of individuals to the mother and thus to national identity. In this regard, Anne McClintock has emphasized that nationalism is constituted from the very beginning as a gendered discourse and cannot be understood without a theory of gender power.[50] Foreigners adopt countries that are not their own and are "naturalized" into the national family. In this way, nations are symbolically figured as domestic genealogies. Calls for a potent nation-state are often understood metaphorically as calls for stable bourgeois families, families in which individuals fill their proper roles for the good of the state. If the nation is the enshrined mother sanctioning this fellowship, then the stability of its proper functioning depends on the degree to which the nation's real women embrace their given identity and enact the destiny of the idealized mother.

As we have seen, the modern, often immigrant woman is compared to a wild mob (in Ramos Mejía), and in Martel she is figured as a prostitute or a diabolical city, but how might a woman author describe the figure of the immigrant? One of the rare positive portrayals of a female immigrant can be found in the best-selling novel *Stella: Novela de costumbres argentinas* (1905), not coincidentally written by a female author, Emma de la Barra, albeit a woman hiding her identity under the male pseudonym César Duayen. By the early 1900s, well-known female writers were not unheard of. These writers never failed to idealize the good wife and mother, with urgent pleas for education close behind, and success in traditional female roles was increasingly linked to a woman's access to education.[51] By the end of 1905, *Stella* was in its seventh printing. Translated into several languages, it had sold more than three hundred thousand copies by 1932, and in 1943 it was made into a movie. Francine Masiello correctly reads the novel against the works of late positivist writers such as Martel.[52] It is perhaps indicative of the social transition taking place in Argentina at the time, in which the masses gained social privileges and destabilized the family, that *Stella* appealed to a variety of readers. While the novel on the one hand defended interests of the oligarchy, it also suggested alternative models of behavior to its audience. The immigrant in this novel seems more a stranger in Simmel's sense than a threatening, degenerating virus. Simmel argues that the stranger is not an irrelevant wanderer but somebody who lives within a group to which he or she did not originally belong. As an outsider, the stranger brings to the group qualities that are not indigenous to it.[53] Not bound by familial roots to the particular constituents and partisan dispositions of the group, the stranger confronts all these with a distinctly "objective" attitude and can therefore be seen as an impartial judge, reinforcing the "true" but corrupted system, or else as an agitator, someone who tries to corrupt the group.

Stella tells of the romance between a cynical middle-aged man, Máximo, and a beautiful Norwegian woman, Alejandra, and is set in the upper echelons of *Porteño* society. Alejandra, whose beauty causes her nieces to become jealous and elicits the admiration of several men, lives with her sister, Stella, an invalid, in their uncle's house after the death of their father. The orphan story thus establishes a female Bildungsroman from the start, inserting his narrative of adolescent growth and development within a paradigm of immigration. Undermining the representation of the unwed woman, who is treated lamentably in social documents and literature of the period, Alejandra, or Alex as she is called, forcefully inserts herself in elite social circles. Unlike her gossipy and unreliable female relatives, she is a woman trained for intellectual life, having been instructed by her scientist father. She learns to think like a man, excelling in the sciences, in mathematics, and in the art of reading. Alex, as family scribe, prepares her Norwegian father's memoirs of his expeditions to the North and South Poles, while at the same time she acts as the record keeper for her family, a generation in decline. Alex saves the family from financial ruin, rescues the patriarch from the abuse of speculative economics, and places the family records in proper order.

De la Barra's insistence on the positive effect of racially appropriate immigration echoes opinions of previous times rather than of her contemporaries. Alex is presented as a role model for Argentine women to follow. For instance, Alex talks about Norway, where women work and where theft is unheard of.[54] De la Barra also integrates questions of women's labor into the question of family survival. In this respect, *Stella* expands the premises of a novel such as Eduarda Mansilla's *El médico de San Luis* (1860), in which the foreigner as family patriarch volunteers to cure the ills of the nation. One can speculate, however, as to why, for her critique of contemporary Argentina, de la Barra chose an immigrant woman. Apparently, by 1905, the figure of the immigrant was already so threatening to the elites in Buenos Aires that de la Barra had to transform the outsider into a woman.

Otherwise, Alex seems curiously reminiscent of the Argentine founding fathers' ideal of the model (albeit male) immigrant. First of all she is racially "impeccable," which comes to the forefront in the first chapter of the novel, when upon arriving in Argentina the women are frightened the first time they see a black child. Being Norwegian, Stella and Alex have never seen a black person before, whereas the Argentine family lives in the proximity of blacks; indeed, La Muschinga, as she is called, seems to be a close companion of the family's children and dances for them upon request, with "the elasticity of an ape."[55] This child, Masiello notices, thus supplies an updated version of the discourse on race: blacks and others, still loathsome, are now reduced to the manageable status of children, scary but harmless. Stella and Alex, however, instinctively feel that "black" equals scary and ugly.

Feared for her foreign observations, Alex becomes the source of national enlightenment, a bridge between tradition and modernity. Ideal immigrant that she is, Alex assimilates and suggests cures for the ills of the nation. As a tutor, Alex points out the weaknesses in Argentine education, described as inadequate for a population still lacking in proper training.

The vocabularies of both kinship and of the home denote something to which one is naturally tied. Since fate lies outside the realm of choice and agency, national identity seems natural.[56] An immigrant, however, has by definition abandoned and betrayed his or her family or country by replacing it with another, and is therefore "unnatural." How can the immigrant Alex then become the heroine in a national romance? In other words, under what circumstances can we empathize with children or citizens who abandon their parents or country? De la Barra skillfully addresses this dilemma, and it is this careful negotiation between nationalist, feminist, and positivist discourses that forms the most challenging aspect of this novel and is probably also a key to its popularity.

Alex cannot fulfill both national and filial obligations at the same time and is forced to choose kinship, albeit reluctantly. Her invalid sister Stella barely survives the passage to Argentina, and a return to Norway would surely kill her. As long as Stella is alive, Alex cannot return to her native country. Effectively trapped in Argentina, taking care of her sister, teaching her cousins, and arranging her uncle's papers—that is, rewriting her Argentine family's history—she also starts writing a biography of her real father, Gustavo Fussler. Thus, the sickly Stella represents the importance of memory. The continuity of a people is composed of its many individual acts of remembering. Alex's writing allows her to remain faithful to her father's Norwegian legacy and at the same time transform this process of mourning into something productive for her host country.

Being among family does not necessarily imply being at home, as Alex quickly finds out. Although they are connected by blood, she seems to belong to a different race than her Argentine relatives. Her family suspects her of being a manhunter and distrusts her "porque no había lazo anterior, no viejo afecto para servir de contrapeso al choque que fatalmente debía producirse" [57] (because there was no prior tie, no old sense of affection to serve as a counterweight to the collision that, fatally, was sure to come). The jealous Clara Montana suspects Alex of desiring an incestuous relationship with her uncle and of trying to seduce him for his money.[58] Even Máximo betrays her when he sees her visit banker Samuel Montana—presumably Jewish—and immediately suspects an affair: "Ya que se la había de comer el cristiano, que se la coma el moro" [59] (The Christian [her uncle] should have eaten her by now, let the Moor eat her now).

The family is in crisis, and interestingly enough, this is due not to the father, but rather to the mother. Twisted mother figures abound in *Stella*, where Alex, by becoming her sister's caretaker, replaces her own mother. The dead mother was the good, albeit uncivilized one, and Alex is clearly her father's child. Don Luis Maura, her older brother, never forgives Norwegian Gustavo Fussller for taking away his beloved sister.[60] Though beautiful, Ana María is uneducated and loves, but does not understand, her husband.[61] Little Alex teaches Ana María history, but her mother likes churches and stories of the saints best.[62] Of all the European countries visited, Ana María likes Spain the best: "la noble tierra de sus antepasados, entre gente de su temperamento, de sus hábitos, de su lengua, entre una raza de su propia raza"[63] (the noble land of her forefathers amid people of her temperament, of her habits, of her language, amid a race of her own race). When Ana María speaks negatively about the Moors, Alex teaches her mother about the Arabs, who were benign and tolerant, allowing the Christians to practice their faith and teaching resignation and poetic melancholy to the Spanish, and by extension the Argentine people.[64] Máximo clarifies the link between Spain and Argentina when he describes his gauchos as "la melancolía árabe vigorizada por el temperamento español. Sarmiento afirma que ha creído ver en África tipos que había conocido en las campañas argentinas. Yo he visto en los bulevares de París un verdadero gaucho con turbante"[65] (the Arab melancholy invigorated by the Spanish temperament. Sarmiento affirms that he believes to have seen in Africa types [of people] that he had already met on Argentine campaigns. I have seen on the boulevards of Paris a true turban-clad gaucho).

Máximo's reference to Sarmiento is, of course, not gratuitous: the latter's disdain for Spain and the Spanish heritage is well documented. Sylvia Molloy has commented on this anti-Hispanic notion and observes that for Sarmiento, Spain represented the bad mother, the source of all the colonies' defects.[66] Yet, at the same time, Sarmiento wrote about a good mother and Spain, represented by the earlier, medieval Spain of the Moors. This Moorish heritage has migrated to Argentina and is, by extension, part of Ana María and Alex's heritage. The unsuitable, "Spanish" mother is the frivolous Aunt Carmen, whom Alex, in an interesting Oedipal enactment, almost replaces, first by becoming a governess to the children and grandchildren and then by helping her sick uncle save the family business. Unable to educate her children, she is concerned with finding them wealthy spouses, is superstitious, and prone to gossip. It is this local influence Alex will have to replace. Immigration policies in *Stella* do not differ very much from the ones promoted in the early nineteenth century. Norway is believed to have culture, while Argentina is not: "Creóse alrededor de las dos hermanas transplantadas por su estraña suerte, de la fría Cristiana a la cálida Pampa, de un medio de refinada intelectualidad a otro medio indígena, de primitiva ignorancia, una atmósfera de devoción y de cariño, y mirábasele como a dos seres de leyenda"[67] (Around the two sisters transplanted by their strange luck, from cold Christiana to the warm Pampa, from an intellectually refined environment to another half indigenous, an atmosphere of devotion and affection was formed, and they looked like two legendary figures). However, there is a clear nostalgia for the Argentine past as indicated by the description of *El Ombú*, the family's country house: "Conservaba la alegría sana, de las sanas generaciones que habían vivido y muerto allí, conociendo como placer el santo amor de la familia, como ley, la santa ley del trabajo"[68] (It maintained the healthy joy, the healthy generations that had lived and died there, understanding pleasure as the sacred love of family, as law, the sacred law of production). It is here, in the countryside, that the family's wealth originated, and it is here that Máximo first starts having feelings for Alex.

By contrast, Máximo's house is full of exotic plants, art, and even electric light. When Alex sees his living room, she is reminded of her father's study: "Estoy impregnada de arte y de recuerdos . . . es éste el primer goce íntimo, espiritual, que he sentido desde que estoy en Buenos Aires"[69] (I am impregnated with art and memories . . . this is the first intimate, spiritual delight that I have felt since I arrived in Buenos Aires). But amid this foreign luxury, they also visit Máximo's labor force, peasants, primitive and courageous, whom Máximo admires.[70]

Don Samuel Montana, a banker with a Jewish sounding name, meets Alex at the horse races and is the main lender to the financially incompetent Uncle Luis. Máximo, fueled by xenophobic prejudice, suspects the two of having an affair, but the banker is an art lover and buys from Alex a valuable painting by the French artist Corot, thus saving the family from financial ruin. And indeed, depressed, cynical, and lazy Máximo suffers from blind xenophobia when he comments to Alex: "Avanzamos por agregación y adopción, lo que nos va quitando todo lo nuestro . . . Los nietos de nuestras grandes familias se substituyen por los inmigrantes, enérgicos y luchadores, pero sin alma nacional, con el patriotismo estrecho vinculado a la prosperidad material únicamente"[71] (We advance by adding and adoption, what continues taking away all that is ours . . . The grandchildren of our great families are replaced by immigrants, energetic and hotheaded, but with no national soul, with a narrow patriotism linked only to their material prosperity).

Alex feels offended by this accusation of lack of patriotism and curtly responds, "Se lamentan de males cuya corrección está en su mano"[72] (They complain of wrongs when they possess the ability to correct with their own hands).

As a surrogate mother, Alex is denied her own sexuality until she can marry Máximo and become the new patriarch's legitimate wife. She succeeds in continuing her father's legacy but will have to sacrifice something else, since, as Bonnie Honig argues, any immigrant will have to sacrifice and mourn that which cannot be carried over from the homeland.[73] In Alex's case, Stella, who contains excellent but unusable qualities for Argentina, namely being a child and an invalid, occupies this role. Her father has named his youngest daughter after the *Estrella Polar*, the name of the boat on which he came to Argentina, and handicapped Stella is compared to a larva but has "nada de enfermizo, nada de morboso en su aspecto, era ella una degeneración, no una degenerada"[74] (nothing sickly, nothing morbid about her, she was unevolved, not a degenerate). Stella is illegible; she does not signify anything, and precisely because of that she comes to signify all the other characters' different, conflicting desires, such as Norway, youth, innocence, the desire to have children, or even the wish to explore the South Pole. Unassimilable Stella is simply too weak to return to Norway or to survive on foreign soil and is limited to mirroring other people's wishes. Not coincidentally, she has her first major stroke just after the celebration of carnival, which was rather raucous in the countryside. Alex turns her father into a type of national heritage for both Argentina and Norway, by writing his biography, and creates out of Stella's death the link that will allow her to marry Máximo and thus truly become an Argentine woman.

CHAPTER 3

NEGOTIATING
NEW IDENTITIES

ARGENTINA OF THE CENTENNIAL

ACCULTURATION VERSUS XENOPHOBIA:
MANUAL DEL EMIGRANTE ITALIANO

FEARS ABOUT THE LOSS OF NATIONAL IDENTITY AND THE idea that Argentines formed a distinctive ethnocultural group that was threatened by foreign influences were constant and pervasive themes of the cultural debates of the Argentine Centennial period. As Carlos Altamirano maintains, a moral crisis emerged at the time of the *Centenario*, which revolved around three principal themes: the racial constitution of the nation, the critique of materialism, and the level of participation in the political process.[1]

Scholarly treatments of this period tend to emphasize the xenophobic nature of Argentine cultural nationalism. Rather than exclusively seeing the ideologies of this period as xenophobic reactions to foreign mobs, I suggest that possible strategies of including these others were just as prevalent. Rather than serving solely as a means of excluding the immigrant from the national community, Argentine cultural nationalism was also propelled by an integrationist impulse.[2] While by no means discounting the Argentine elite's fear of social upheaval and its distaste for the working class immigrant, this line of interpretation seems only partly convincing. As I argue, cultural nationalism[3] was not so much an attempt to reject the immigrant as it was a means of integrating the newcomer into the national community in a way that marginalized him politically. While deploring the newcomers as a threat to the collective race or character, cultural nationalists and their sympathizers accepted, albeit at times begrudgingly, that immigration was inevitable and believed that the incoming masses should be assimilated or "Argentinized" as completely as possible. If, during the early decades of the twentieth century, growing numbers of native Argentines began to understand their nation as a unique ethnocultural community and saw themselves as forming a distinctive race, what role did they envision for the millions of immigrants flooding their shores? Could the immigrant become a member of the Argentine race, and if so, how was this to be accomplished? The emergence of an ethnocultural understanding of "Argentine nationhood" coincided with, and indeed was in large part precipitated by, a massive influx of European immigrants.

One of the most important changes in Argentine thought, frequently overlooked, is the concept of nation itself. During much of the nineteenth century, liberal Argentines—inspired by France's example—had understood their nation to be a political association, open to all who embraced a common political creed and worked for the welfare of the nation. By the opening decades of the twentieth century, however, a significant group of young intellectuals, known in Argentine historiography as cultural nationalists, began to espouse a vision of the nation that more closely resembled the ethnocultural conception of nationality. In this new interpretation, the Argentine nation became a "motherland," or patria. This ethnocultural vision of the nation, rather than providing a rationale for marginalizing the foreigner, actually served as a means of integrating him. The emerging "raza argentina," then, would include rather than exclude the immigrant masses, who should be in time assimilated or "Argentinized" as completely as possible.

Economically, immigrants proved quite successful, inserting themselves in a position above the unskilled Argentine masses but below the traditional landed elite. As studied by Samuel Baily, mutual aid societies became an important factor in organizing mainly Italian immigrant groups upon arrival.[4] Most of these societies were founded during the last two decades of the nineteenth century, and a number of large ones (with one thousand or more members) dominated the movement and provided a distinct kind of leadership. These societies had substantial assets in both buildings and capital reserves. They also performed more extensive services for their members. In addition to the normal insurance and social benefits, they provided schools, medical clinics, hospital care, pharmacies, restaurants, and, in some cases, job placement services. Although skilled artisans formed the largest subgroup of members in most Argentine societies, these organizations also included substantial numbers of semiskilled and white-collar workers. Italians from all parts of the peninsula joined these societies and rose to positions of leadership within them. The large societies were able to develop a relatively united mutualist movement before World War I.

Baily also mentions the importance of Italian-language newspapers, read to find job opportunities, activities of community organizers, tips on how to survive and prosper, guidance in understanding and adjusting to the wider community in which they lived, and information on Italy. *La Patria*, the first Italian newspaper, was established in the 1860s and was the most widely read of this period. Its tone was generally liberal and anticlerical, regularly publishing information on working and living conditions and examining grievances of workers involved in labor disputes.

Apart from the foreign language press, conduct manuals were regularly to inform the immigrants of basic facts and customs of their host country. Sometimes manuals were commercial, sometimes meant for a very limited region, sometimes free, and they were often read aloud. Diego Armus has translated and published an abbreviated version of the *Manuale dello emigrante italiano all'Argentina*,[5] originally written by Arrigo De Zettiry, a functionary of the Reale Commissariato dell'Emigrazione, who published this in 1913 by the Reale Commissariato dell Emigrazione in Rome. This manual presents the names of major shipping companies, some Catholic charity orders, and official agencies related to immigration. Other topics vary from legal formalities to the Italian immigrant's right to purchase products from his or her home country. Railway stations, useful addresses, a small Spanish-to-Italian dictionary, tourist attractions, history, and geography are all presented as if in a school textbook. There is a brief outline of the basic laws, including the updated one on military service, described as a small contribution to the host country that does not require one to betray his country of origin. Although the city was the main destination of the majority of people, it barely figures in this manual. There is no mention of the conditions on the ship or the exploitation in the countryside. The readers are urged to see some of the tourist attractions but then leave as soon as possible for the pampas. Official buildings in Buenos Aires are mentioned, but not the *conventillos* or tenement buildings where most newcomers would end up. There is a brief mention of the Hotel de inmigrantes—recently updated according to the latest hygienic standards—but the newcomers are urged to eat with knife and fork and be careful with their table manners. It is also very important, De Zettiry cautions, that ladies take off their hats in the cinema, so they do not block the view of others, and never to address a lady as a "donna." One wonders why the manual is so riddled with trivia, while ignoring the anxieties and discrimination that many immigrants doubtlessly experienced. References to issues like poverty or possible political affiliations are also absent.

Since the majority of Italian immigrants came from rural areas, it is not surprising that they should initially dream about having their own land. In his prologue, Armus makes an interesting comment about the pastoral utopia that these manuals create:

[Estas] guías descubren sugerentes singularidades de la historia íntima de 'hacer la América.' Las de 1870 hablan de 'tierra prometida,' de ventajas y posibilidades ilimitadas, de colonización, de seguro acceso a la propiedad de la tierra, de fortunas fáciles . . . Comenzaron entonces, a escucharse las primeras voces de una peculiar utopía agraria, al tiempo que las facilidades ofrecidas por los países de recepción—desde los terrenos a las herramientas y animales- incitan a muchos a la experiencia migratoria. Eran los colores de un cuadro casi paradisíaco y de una tierra increíblemente rica los que convocaban las fantasías e ilusiones de pobres y pequeños propietarios esperanzados en una vida distinta.[6]

[(These) guides reveal suggestive peculiarities of the intimate history of "making America." Those of 1870 speak of a "Promised Land," of advantages and unlimited possibilities, of colonization, of

sure access to land ownership, to easy riches . . . Then, we began to
hear the first voices of a peculiar agrarian utopia, while the facilities
offered by the receiving countries—from the plots of land to tools
and animals—incite many to immigrate. The colors of an almost
paradisiacal painting and of an incredibly rich land fueled the fan-
tasies and dreams of the poor and of the small landowners hoping
for a new life.]

De Zettiry persists in talking about the colonization of empty,
virginal territories when they were really referring to manual labor
in rural areas. Again, one is left wondering why this guide seems
so careless about the interests of those it is supposed to help. It is
very possible that a bureaucrat like De Zettiry never traveled to the
interior and simply repeated the old clichés of "hacer la América."
Still, it is interesting to note that somebody like Alberto Gerchunoff—who definitely had the Jewish colonists' interests at heart—
also insists on this agrarian utopia.

It almost seems as if the "ideal Italian immigrants" De Zettiry
had in mind were already the middle class; assimilated, hard-work-
ing people that Argentina was supposedly looking for, conveniently
located somewhere in the countryside, politically inactive, and with
sons eagerly fulfilling their duties in the Argentine military. Disregarding the rural origins of the vast majority of immigrants, the
manual seems directed to a middle-class audience, which makes
one wonder if, rather than informing those who most needed it,
the manual was more directed toward improving the stereotypical
ideas urban middle-class Argentines had of these newcomers. The
condescending attitude of Italian officials throughout the work is
striking. Although published in 1913, long after the first waves
of mass immigration, there is no mention of the previous immigrants, Italian or otherwise. It is as if they never existed or had
blended in with Argentine landowners. Nor does the manual ever
mention with any specificity the different regions of origin; rather,
all Italian emigrants are transformed into a shapeless mass, driven
along by "destiny."

THE CENTENNIAL PROJECT OF THE
NATIONAL INTELLECTUAL

It was widely held by the liberal oligarchy that in Argentina a new race was forming, one that would represent an amalgam of the diverse racial groups, including the immigrant masses. Consequently, the notion of a national heritage and culture was of crucial importance. Designed to commemorate the independence from Spain in 1810, the goal of Argentine Centennial intellectual projects was to promote a nationalist ideology celebrating "traditional" values. This reassessment took several forms—the revival of interest in national history, nostalgia for Hispanic cultural roots, insistence on Spanish as the language of education, and even the rehabilitation of the Indian as a romantic element of Argentina's past.

As Beatriz Sarlo and Carlos Altamirano point out, this is exactly the period when the concept of a public intellectual also changes. Writers are increasingly invested with "antimaterialistic" and "spiritual" values, and only they possessed the moral purity that allowed them to understand the true nature of Argentina's cultural essence. Turn-of-the-century intellectuals still saw the immigrant as an agent of "modernity" and the gaucho as the unique timeless creature of the pampa—unsuitable for economic progress. But the progress of modernization itself had been called into question.[7] The most significant reaction to the generation of 1880 and the liberal state program was manifest in the debate about "identidad nacional" (national identity) and the development of "cultural nationalism," most notably in the writings of Rojas and others who were active in intellectual circles in the early 1900s.[8] As writing became a viable profession toward the end of the century, and as literacy rates began to increase, intellectual and political elites began to divide. Having lost their traditional status as shapers of national affairs, many early twentieth-century intellectuals sought to establish a new role for themselves as the guardians of authentic Argentine cultural traditions and values.[9] As William Acree points out, these intellectuals were more than just guardians of the nation. In addition to the "problems" of modernization and immigration, intellectuals and politicians alike worried about the tensions generated with Argentina's entrance into the capitalist world economy: labor strikes, protests, the spread of socialist ideas, the possibility of anarchy, and so on. Faced with these questions, intellectuals like Rojas set out to find, redefine, or reinvent the *alma nacional* (national soul).

The primary task of the "new intellectual" was that of discovering a true national identity, of developing a cultural nationalism. Politicians asked writers to develop a modern cultural program that would bring prestige to Argentina.[10] The first Argentine literary history and the debates about the importance of the epic poem *Martín Fierro* are clearly connected to this nationalist project. Under the term *la argentinidad* ("Argentineness"), coined by Ricardo Rojas in *La restauración nacionalista*, writers started to look for the "essence" of their country. Their texts reflect the writer as a guide, an advisor of a patriotic project, and the historical interests of the Centennial project thus posited the writer as a cultural interpreter. Not coincidentally, the first chair of Argentine literature, filled by Rojas, was created in 1912 at the University of Buenos Aires. The growing importance of spiritual philosophies, a direct idealist reaction against positivist thought, influenced by Nietzsche's idea of a civilization of only superior beings, was quite in vogue. Another important element in Centennial thought was the changing attitude toward Spain after 1898. The loss of Spain's last colonies and the growing importance of the United States encouraged a revision of history and a focus on the "Hispanic race." Argentine intellectuals such as Manuel Gálvez and Rojas read Spanish nationalists Miguel de Unamuno and Ángel Ganivet and wanted a rebirth of the national soul.[11]

However, it is important to keep in mind that this new Centennial spirit was partly a continuation of nineteenth-century liberal policies, since the goal was still to create a modern, stable, homogenous state from above; the majority of the inhabitants were seen as cultureless barbarians. Even as a peripheral, dependent state, Argentina did experience an economic boom and with that, urbanization and class conflicts, both from anarchist and socialist groups as well as from rising middle classes that aspired to more prominent political positions.[12]

Given this economic success and the fact that immigrants formed both the core of Argentina's urban working class and the emerging middle or entrepreneurial class, few immigrants demonstrated any inclination to integrate politically. Naturalization rates were surprisingly low during this period; only 2 to 3 percent of all immigrants to Argentina became citizens. Francis Korn provides the following statistics: naturalization in Buenos Aires in 1895 was 0.2 percent and in 1914 only 2.3 percent. In the United States, by contrast, naturalization percentages were 63 percent in 1890 and 65.2 percent in 1900.[13] Korn suggests that so few immigrants bothered to become naturalized citizens because assimilation was much easier in Argentina than in the United States. Immigrants did not feel the need to be naturalized, owning to a greater degree of acceptance, and were not afraid of being sent away, since racial quotas such as those used in the United States did not exist in Argentina.[14] Ghettos did not exist in Buenos Aires and immigrants were never physically isolated. Halperín Donghi correctly points out that Sarmiento, who, as mentioned earlier, became so disappointed in the civilizing mission he expected immigrants to fill, did not stop to question why immigrants were so reluctant to adopt Argentine nationality and thus remained unable to vote. For the successful immigrants, the new political system was ineffective, whereas the poor ones preferred to retain the protective laws of their country of origin that were designed to help them.[15] What unites Sarmiento and the cultural nationalists is that their philosophies are born out of concern for the same well-known dichotomy—civilization versus barbarism—and the ideological positions they defend are closely connected, through both the integration of this dichotomy in their visions of Argentina and their reactions to previous national ideologies.[16]

It is worth asking, however, why so many prominent intellectuals of the period completely ignored the naturalization issue and why they had so little faith in the political arena as a source of national cohesion. It would be false to portray all intellectuals who embraced the ethnocultural understanding of nationality as antidemocratic. Ricardo Rojas, despite his obvious elitism, remained a supporter of democratic institutions. An implacable critic of the military coup of 1930 lead by José Félix Uriburu and the government that followed, Rojas endured two years of internal exile in Tierra del Fuego for his beliefs. The question is then why did Rojas and others with democratic inclinations not see formal citizenship as an essential component of Argentineness? Why did they not see a common political life and uniform citizenship rights as the basis of national solidarity and cohesion or view bringing the immigrant into the political process as a way to strengthen the nation? This is an issue of much debate among scholars.[17] One of the most convincing hypotheses is suggested by Hilda Sábato, according to whom early twentieth-century Argentines had had very little experience living in a functioning democracy. Despite the ideals of nineteenth-century leaders and the 1914 electoral reforms, democracy had very shallow roots in Argentina, and the notion of citizenship was extraordinarily weak. Politics had always served to divide rather than unite the nation, with factional disputes all-too-often resolved with bullets rather than ballots. Given this history, it was not surprising that few intellectuals would see the political sphere as hospitable to cohesion.

Sábato is of course writing about the period before the rise of cultural nationalism. While her argument certainly seems plausible, I suspect that once again, the change in the concept of "nation" might also explain the cultural nationalists' lack of interest in the naturalization of recent immigrants. When nationality becomes conflated with ethnicity, and membership in the national community is a question of descent rather than consent or territorial residence, with this understanding of nationhood, people cannot choose or acquire their nationality, which would make the issue of nationalization obsolete. As Delaney points out, the properties through which a people becomes a nation concern the collectivity and national character, perceived as a prepolitical essence.[18] What had changed since the mid-nineteenth century was not simply the assessment of the immigrant but also the underlying understanding of what nations were and what held them together. Alberdi and Sarmiento thought that native Argentines could acquire, through simple contact or more formal education, desired Anglo-Saxon traits. This desired Anglo-Saxon immigration was not supposed to improve the "genetic stock" of the country but to help transform the work habits and customs of this native population. As mentioned previously, the meaning of "nation" had gradually moved from a voluntary association to an ethnic concept of a new national race and character. Cultural nationalists resolved this dilemma by focusing on supposedly unique cultural and historical elements rather than biological traits.

EDUCATION AND CITIZENSHIP: RICARDO ROJAS

Foreign forces that threatened the essence and traditions of Argentina served to invert Sarmiento's nineteenth-century civilization-barbarism dichotomy and to replace it with one that pitted the authentic or invisible Argentina against the visible, or inauthentic, Argentina.[19] Perhaps the most representative intellectual of this period who sought to create a nationalist culture as a defense against Argentina's threatening cosmopolitism was Ricardo Rojas. In 1910, the Argentine state assumed control over the country's educational system. One year before, Ricardo Rojas published *La restauración nacionalista*, a report on the state of education requested by the government with the purpose of "correcting" the content of the civilizing mission of education. Born in Santiago de Estero, a descendant of one of Argentina's original settlers, he initially published a series of articles in *La nación* attacking the Argentine intelligentsia's blind acceptance of cosmopolitanism and expounding the view that the greatest contribution to the development of the Argentine nation had been made by the Hispanic Argentines.[20] Rojas underscored the importance of studying Argentine history and the uniqueness of Argentine traditions to form a national "consciousness." He was particularly concerned with the need to emphasize the republic's historical tradition within the national school system. Reporting to the Minister of Education in 1909, Rojas suggested that an awareness of the country's history would serve to form a "national spirit" and to "Argentinize" the children of European immigrants who would be entering Argentine schools. Rojas perceived the uniqueness of the Argentine experience to be the mixture of native and European cultural traditions on American soil. In his writings he urged Argentines to cease imitating European cultural standards or reverting to a reactionary nativism and instead to fuse the best of both in a national synthesis. The importance of idealism as opposed to the positivistic and materialistic values of late nineteenth-century Argentina was closely related to the Argentine intellectual's search for a unique national identity.

Rojas wrote that nationality and nationalism had replaced "patria" and patriotism, that in the coming age only those groups with a collective conscience would survive and prosper, and that

Argentina should aspire to be an ethnic and spiritual entity rather than simply a political one. He criticizes Buenos Aires as a city where signs in Italian or Yiddish were displayed in the shop windows of many traditional and previously criollo neighborhoods and where children of immigrant origin were mixed with the old Hispanic population, thus endangering linguistic purity. The immigrants were becoming the new barbarians, the unwanted result of the policies that the elite of the last third of the nineteenth century had executed following the enlightened program of Sarmiento and Alberdi, in order to build through state action a modern capitalist society. In fact, the outcome of the modernist project was considered a distortion, the unforeseen consequence that transformed Buenos Aires into not only a cosmopolitan city but also a dangerously mixed space haunted by cultural loss and political turmoil led by anarchists and socialists of foreign origin.

A nationalist conservative, Rojas held a strongly negative view of the effects of the country's modernization on its spiritual life. His writings tended to concentrate on the cultural incompatibilities of certain "races" and the Argentine Indo-Hispanic heritage, disregarding the modern, progressive, and "scientific" approach of the liberal intelligentsia. The rejection of materialism and the need to cultivate artistic values, beauty, and idealism were, of course, key tenets of the modernist movement that swept Hispanic America during this period and gained numerous followers in Argentina. *Modernismo*, with its hostility to positivism, clearly nourished and helped provide the vocabulary for many intellectuals' critiques of materialism.[21] According to David Rock, Rojas, too, conflated "national character" with federalism and its political leaders, believing that the so-called age of barbarism, so reviled by Sarmiento, was actually the most genuine source of Argentine character.[22]

A positive assessment of this Latin identity was under way in opposition to the "marginal" others migrating to the capital. Carl Solberg has pointed out that the centenary celebrations, showpieces for invited foreign dignitaries and the fashioning and promulgation of Argentine culture, were themselves intertwined with the anti-immigrant violence of the period, in an attempt to suppress the growing influence of these newcomers.[23]

While avant-garde writers and intellectuals sought to read and translate high European literature for the sake of bettering Argentine letters, they also rejected the languages of the immigrant masses and their print manifestations in newspapers and popular literature. Fears about immigrants as avatars of radical, foreign ideologies and as carriers of an inferior race contributed greatly to the anti-immigrant sentiment sweeping Argentina at the turn of the century. Yet another element was the fear that the immigrant was undermining national identity, diluting a specifically Argentine "character." The cultural nationalists' view of the Argentine nation as a unique ethnocultural community seems to be strikingly at odds with the Sarmiento of *Facundo*. The distance between the nineteenth-century vision of the immigrant as a source of democratic values and the early twentieth-century view of the immigrant as an agent of national dissolution was great indeed. Still, I disagree with Diana Sorenson's hypothesis that Rojas directly confronts Sarmiento in his writings.[24] It is important to remember, as Adriana Puiggrós points out, that for Argentine elites, Sarmiento's words were practically a mandate, and to ignore him would be to return either to barbarian "chaos" or to colonial Catholic conservatism. However, traditional liberal thought had provided the immigrant with a key role in the national pedagogical project as educators who would bring civilization to the country. In the Conservative Republic, however, this same immigrant had become "barbaric." In the early 1900s, the ruling elites had to find a balance between Sarmiento's inherited convictions and safekeeping and policing national culture. As Puiggrós explains, Sarmiento's discourses were read in a closed, ritualized fashion dictating the "truth."[25] Fortunately for them, however, and Sorenson completely disregards this aspect of Sarmiento's writings, Sarmiento is of course notoriously contradictory and his writings easily lend themselves to conflicting interpretations. Another factor, also unmentioned by Sorenson, which helped the cultural nationalists' reinterpretation of Sarmiento's legacy, was the change in the author himself, who over the years had grown considerably more pessimistic over the immigrants' supposedly beneficial impact.

In his introduction to *La restauración nacionalista*, Rojas places himself as a direct inheritor of Sarmiento.[26] Glossing over the obvious inconsistencies in the polarization between "civilización" and "barbarie," Rojas simply states that Sarmiento "tuvo la suficiente libertad mental y acierto político para censurarlos en el 'gringo' cuando la emigración ya realizada planteó nuevos problemas morales a la nacionalidad argentina"[27] (had enough mental freedom and political intelligence to condemn the "foreigner" when the emigration that had already taken place posed new moral problems for Argentine nationality). Rojas then implicitly draws a distinction between cosmopolitanism and true enlightenment, trying to pass this off as Sarmiento's idea:

> El cosmopolitismo es una forma de barbarie . . . no es internacionalismo ni humanitarismo; es anarquía espiritual de una sociedad, comporta el empobrecimiento de sus fuerzas históricas. Pueblo que aspira a realizar una obra de cultura, debe superar el cosmopolitismo por un ideal nacional. El nacionalismo en los países de necesaria inmigración como la nuestra, es una disciplina idealista en defensa de la civilización.[28]

> [Cosmopolitism is a form of barbarism . . . it is neither internationalism nor humanitarianism; it is the spiritual anarchy of a society, it brings the impoverishment of its historical forces. A people that aspires to complete a work of culture, should overcome cosmopolitanism for a national idea. Nationalism in necessarily immigrant countries such as ours, is an idealist discipline in defense of civilization.]

For these cultural nationalists, "Argentina" denoted a people bound together by common historical memory, language, shared mental and emotional traits, and they had nothing to do with the emerging science of genetics and heredity. For instance, the poet Leopoldo Lugones draws a direct relationship between mixed races, mixed languages, and anarchy: "Given the inferior social condition of cosmopolitan immigrants, they tend to deform our language with pernicious contributions. This is extremely serious since that is how the disintegration of the fatherland begins. In that respect, the legend of Babel is very significant: the dispersion of men originated with language's anarchy."[29]

The figure of the gaucho, however, acquired new cultural meanings in the 1880s, in large part because of the reactions to the massive numbers of immigrants arriving to Buenos Aires. Delaney argues that the inflow of immigrants, coupled with the new financial connections to Great Britain and the ways in which modernization occurred, provoked the widespread belief that the path to national renewal lay in a return to the values of the Argentine gaucho.[30]

Rojas played the major role in introducing cultural nationalist thought to Argentina. After writing articles on patriotism and nationalism for the newspaper *La nación*, he published these ideas in 1909 as *La restauración nacionalista*. As much a nationalist manifesto as an educational report, his thesis was that Argentina, faced with the threat of invasion by foreign influence and ideas, needed to develop a collective conscience based on its own traditions. This could best be accomplished through the teaching of history in public schools. Also, Argentine geography, language, folklore, and civics should be taught to help awaken a consciousness of the nationality and the spirit of nationalism. Rojas also believed the foreigners' contributions to this developing national personality should be tightly supervised. His well-known activities in the realm of education amply reflect this conviction, proclaiming the need to transform the nation's schools into the "hearth of citizenship" and calling for a complete reorganization of the national school curriculum. This new curriculum should focus on Argentine history, the Spanish language, Argentine literature, Argentine geography, and moral instruction and should seek to inculcate in all immigrant children a love for the nation and an understanding of Argentine traditions. Public schools, Rojas believed, should be instrumental in the effort to "define the national conscience" and bring about a "real and fecund patriotism." In other words, the emergence of the new Argentine race, while gradual, would not and should not be allowed to occur naturally. Rather, to safeguard authentic Argentine values and traditions, the artistic and intellectual elite (individuals such as Rojas) should direct and shape the personality of this emerging race. Rojas believed it possible for the state to teach immigrant children to become Argentines and that Argentineness was acquirable—that is, more a matter of conscious choice that an immutable state of being.

La restauración nacionalista was designed to contribute to an ongoing debate on education, which Rojas and others viewed as the chief instrument to achieve the assimilation of immigrants. Cosmopolitanism, Rojas declared, had destroyed the country's "moral unity," producing an unstable society because its population remained rootless. To rebuild a sense of community, "we must renew our history, cultivate our own legends, revive an awareness of tradition," because "from a sense of its past a people develops a more powerful commitment to its self-perpetuation."[31] Therefore, it was now time to impose a nationalist character on Argentine education through the study of history and the humanities. For Rojas, cosmopolitanism, the dissolution of the old moral nucleus, was seen as the increasing decay of tradition, the corruption of language, and the unscrupulous pursuit of wealth. In Argentina, Rojas argued, "telluric forces" had a unifying function, serving to transform or nationalize the millions of foreigners who continued to pour onto national shores. Rojas's mystical musings must be seen as an attempt to reconcile the contradictions between the ethnocultural understanding of nationality he embraced and the realities of early twentieth-century Argentina. By arguing that these telluric forces of the Argentine soil would impose a common mental or spiritual matrix on the newcomers, Rojas is able to stretch the parameters of the ethnic vision of nationality to make it capacious enough to accommodate the immigrant.

Yet, despite his appeals for educational reforms geared toward instilling patriotism in the school-age population, Rojas never strayed far from the Romantic vision of nationality. For him, becoming an Argentine was always much more than simply embracing a political creed, gaining knowledge of Argentine culture and traditions, or speaking Spanish. Even in *La restauración nacionalista*, in which he outlined his educational proposals, he suggests that civic instruction had its limits. The true nationalizing force, he believes, is the telluric power of the Argentine territory. What will endure are the children of the immigrants and their descendants.[32] Education is crucial: only through the study of Argentine history will the foreign masses become true national subjects. The error Founding Fathers Sarmiento and Alberdi made, Rojas claims, was to confuse material progress with spiritual civilization.

Now the oligarchy could have eliminated educational possibilities altogether, but this would have been ideologically impossible, with the insistence on education as one of the cornerstones of their thought. There were attempts to "modernize" education by stimulating technical and professional careers, but with the rise of Hipólito Yrigoyen to the presidency, these reforms were immediately annulled. In his study on Argentine education, Juan Tedesco argues that the policy in favor of diversification of education appears to be democratic, but its hidden agenda was to prolong social differences and internal divisions. It was said that humanistic values represented the national character, whereas technical and professional careers promoted utilitarian thought, which was inherently *extranjerizante* (foreignizing).[33]

Adriana Puiggrós also convincingly argues that the popular groups, socialist and radical, were strong defenders of Sarmiento's pedagogical values, even though they rejected the government's conservatism. They promoted the development of civil society and public education, for, as she explains, even though the actors had changed, the values had not.[34] Eventually, the modern school system was one of the pillars of the state, organized according to the bourgeois wishes confronting the traditional oligarchy. Through his journalism and his academic work, Rojas aspired to show how the national spirit had manifested itself in the past so that future generations might be guided along the correct path.

THE LANGUAGE OF THE ARGENTINES

Beatriz Sarlo discusses how the process of modernization and changes in the urban landscape of Buenos Aires affected how many understood the idea of nation: "No se habla de otra cosa en muchos de los ensayos producidos alrededor del Centenario . . . El impacto de la transformación no era sólo ideológico; los cambios eran un hecho irreversible y la inmigración ya casi había concluido su tarea de convertir a Buenos Aires en una ciudad de mezcla"[35] (No one spoke of anything else in many of the essays produced around the Centenary . . . The impact of the transformation was not only ideological but also irreversible, and immigration had almost concluded its task of converting Buenos Aires into a mixed city). Differences over the content of the emerging national race or personality are also evident in the growing debate over the national language. Did Argentina need its own national language in order to be a real nation, or was Spanish the true national language? The dilemma for cultural nationalists and their sympathizers was not merely theoretical, for many feared that Argentine Spanish was changing. During the early years of the century, two distinct *jergas* or jargons, both associated with the working-class immigrant population, emerged in Buenos Aires. *Lunfardo*, an urban street slang with heavy Italian influence, and *cocoliche*, a kind of gaucho-talk associated with popular theater that featured dramatic comedies about rural life, were very much in vogue among the immigrant working class. While a few intellectuals applauded these new jargons as evidence that Argentina was at last developing its own language, others ardently defended pure Spanish.

Against the increasing changes in Argentine Spanish caused by foreign-born speakers and their offspring, Argentine nationalists had to develop local, "autochthonous" traditions. Beatriz Sarlo points out that during the 1920s and 1930s, social conflicts extended to cultural and aesthetic debates, and first among these was the question of language, of who spoke and wrote an "acceptable" language.[36] What scandalized or terrified the nationalists influenced the vision of the intellectuals of the 1920s and 1930s. Writers discovered that there are two types of foreign languages, on the one hand those written and read by intellectuals and on the other hand those written and spoken by the immigrants.

Rather than assuming that immigrant groups were passively waiting to be accepted into Argentine society, a *crisol de razas* (crucible of races), it is important to examine some of the ways they found to insert themselves within this nationalist framework. It would be impossible to assume that there would be only one way in which this would happen; in fact, there were as many strategies as there were immigrant groups, highly diverse and with varying degrees of success. I will subsequently analyze the way in which Alberto Gerchunoff claims a "gaucho" identity for his Jewish settlers in *Los gauchos judíos* by using much of the same rhetoric as the cultural nationalists used.

ALBERTO GERCHUNOFF'S JEWISH GAUCHOS

According to Víctor Mirelman, 1890 marks the beginning of Jewish mass immigration to the country. During 1891, the Moroccan Jews founded their first institution, *Congregación Israelita Latina* (Latin Israelite Congregation), while the Russian Jews founded their first society, Poale Zedek (Israelite Worker's Society). The Jewish population in Buenos Aires was cosmopolitan, with people arriving from Poland, Russia, Romania, Syria, Turkey, and Morocco.[37] Their Jewish identity was reflected in different ways, depending on their previous traditions, with Ashkenazim Jews comprising four-fifths of the total Jewish population. They founded tens of religious, Zionist, welfare, cultural, and educational institutions.

One of the most unusual texts to be published around the Centennial was that of the Russian-born Alberto Gerchunoff (1883–1950), whose *Los gauchos judíos* came out in 1910. Before *Los gauchos judíos*, diaries and letters of members of Jewish immigrant communities had been published in Argentina, but Gerchunoff's text became one of the most popular of its time and eventually gained canonical status. The history of the Eastern European Jewry in Argentina is deeply marked by utopian Zionist ideologies. The Bavarian-born Maurice de Hirsch, to whom Gerchunoff dedicated this work, was most responsible for the promotion of immigration to Argentina. Concerned over the renewed anti-Jewish activities in Russia beginning in 1881, the baron arrived at the conclusion that the Argentine plains would offer not only a refuge from pogroms but also, eventually, a new Jewish homeland. The many variants of Zionism in nineteenth-century Jewish thought were far from unanimous in urging the return to Palestine: Hirsch supported the effort in Palestine but believed that Zion could be equally well established in Argentina. There were many projects, some of which were never completed, but others, at least for a while, were quite large.[38]

Originating in Proskuroff, Russia, Gerchunoff's family had first been assigned to Moisésville in one of the Hirsch's colonies, later moving to Rajil in the province of Entre Ríos. After the murder of his father, the family moved to Buenos Aires. In 1908, Gerchunoff was admitted to the writing staff of the elite newspaper *La nación* in which he started publishing *Los gauchos judíos*. He was called on to write the official Centennial portrait of the Argentine Jewry by his friend Leopoldo Lugones; the result was *Los gauchos judíos*.[39] This episodic narrative was composed of nearly two dozen vignettes. Its protagonists, Eastern European Jewish immigrants, find harmony with the rural environment and its inhabitants, the gauchos.

Gerchunoff transfers Jewish immigrant culture to the nationalist nostalgia of the early 1900s in Argentina and marks an era of transition from an Argentina of elites to an Argentina of the masses. Both an immigrant and a member of nationalist avantgarde circles, he attempted to bridge the chasm between upperclass and immigrant formulations of Argentineness. Gerchunoff's work has not aged well and is occasionally incredibly irritating precisely because he is so skilled at reconciling an immigrant utopia with Argentine nationalism. *Los gauchos judíos* clearly presents an urban dweller's view of what an idyllic agrarian community might look like. There are no "hard facts" mentioned, such as the number of people who did not survive the colonies, or decided to leave, or the amount of children who were born. Naomi Lindstrom accuses Gerchunoff of betraying his own project but is unwilling to recognize that betrayal is an unavoidable element of this project.[40] How does Gerchunoff's perspective on Spanish culture, as part of the *hispanismo* movement, relate to his participation in the construction of a particular type of Argentine national subject?

Rather than proving whether Gerchunoff was historically "right" or "wrong" in his portrayal of the immigrant colonies, I am more interested in examining how the pairing of two figures—the Jew and the gaucho—is used rhetorically to negotiate both conservative nationalist and immigrant identities. I suggest that his main strategy is the description of an agrarian utopia. In this utopia, three things stand out: the strictly rural environment, the assimilation process of women, and the negotiation of linguistic and cultural differences. *Los gauchos judíos* should be seen as an attempt to "translate" one culture into another, albeit through highly asymmetrical power relations.

According to Senkman, despite its very positive initial reception, the work was never reedited. The one reprint during the author's lifetime was the second corrected edition of 1936. Apparently, Gerchunoff felt so dissatisfied that he refused a prize offered by *La nación* in 1940 to commemorate the thirtieth anniversary of its initial publication. It was only after his death that the book was printed again. In the early 1970s, there was a rather disappointing movie made by Juan José Jusid, and apparently a comic book version and children's adaptation are now available.

Los gauchos judíos opens in a pastoral biblical tradition, with quotations from the Old Testament.[41] A brief prologue praising Argentina follows: the republic is celebrating its greatest festival—the glorious feast of its liberation. The occasion is compared to Passover; this is a greater Passover than the ones celebrated in Russia. The prologue is dated Year of the First Argentine Centenary.[42] The first story describes the decision of the Jews to emigrate. Tulchin, their hometown, is dreary, and anti-Semitism is rampant. The Cossacks burn the sacred books, so the Jews decide to emigrate. A representative consults with representatives of Baron de Hirsch about the organization of Jewish colonies in Argentina. They discuss Spain as another possibility, and one rabbi calls it a wonderful country if it were not for the curse of the synagogue that still lingers over it. Another rabbi reacts indignantly and curses Spain since it was there that "our brothers [were tortured] and . . . our rabbis [burned], but also because there, the Jews left off cultivating the land and growing livestock. In the first book on the Talmud, farm life is referred to as the only healthy one, the only life worthy of God's grace."[43] Eventually, they decide to immigrate to Argentina, "and we'll go back to working the land and growing the livestock that the Most High will bless . . . If we return to that life, we will be going back to our old mode of life, our true one!"[44]

Three geographic entities are referred to as Jewish spaces, in ascending hierarchy: Russia, Spain, and Argentina. Russia, the place of origin of the immigrants, is repudiated from the very first paragraphs for its persecution of Jews. The dean of the rabbis curses Spain not only for having victimized his people but also for having been the place where Jews abandoned working the land. Argentina, where Baron de Hirsch was establishing the Jewish colonization association, then, becomes the land of salvation, the new Zion.

Favel Duglach, the protagonist of "The Poet," is paradigmatic of the new Argentine Jewish identity. Although he is a lazy farmer, weak, and thin, with skin as yellow as a flame, he is deeply religious, possesses a thorough knowledge of the Sacred Works as well as Jewish legends, which he embellishes in his own way. His stories, even more than his preference for gaucho clothes, allow him to become a true Argentine gaucho: "[He] could feel the native Argentine epics of bravery with the same exaltation he experienced when telling some story from the Bible to a tense, expectant group in the synagogue."[45] Favel is thus described as a humble farmer, a strict traditionalist, a mysteriously terse Other, and a cultured, refined poet. Although he is not one of the handsome immigrants Sarmiento and Alberdi dreamt of, he can become an Argentine by virtue of his creativity. He understands the spirit of the gaucho epic and has words of praise for the well-plowed field. As Francine Masiello observes, the return to the "autochthonous" roots of the countryside allow Centennial writers to ignore contemporary problems and hide themselves in a privileged world, free of destabilizing threats of disorder.[46] Gerchunoff eagerly uses this technique. The process of Jewish immigration and acculturation is thus carefully located in the countryside, far away from the dangerous city. It is also presented as a process specified not in a concrete historical time, but rather in the realm of the epic birth of the Argentine spirit itself.

The idea of the corrupt city is reinforced in Favel's tale that Gerchunoff narrates in some detail. In Babylon, where captive Jews prayed to Jehovah in spite of the naked courtesans dancing in front of the temple, one day a mysterious handsome Jew arrives and organizes the captives to fight for their freedom.[47] Unfortunately, they lose, and the handsome young man is tortured to death. Although the theme of the messianic savior is obvious, the topic of the decadent city one has to escape is also immediately introduced in this story. Dalia Kandiyoti has rightly remarked that, while Argentine nativism finds its personification in the countryside's vanishing gaucho with his reinvented virtues and intrepid mode of life, Gerchunoff's new immigrant Jews, who could not resort to such an identification, reinstate another tradition on the very site of Argentine pride: a picturesque, biblical way of living, a timeless Jewish tradition continued on benevolent Argentine soil.[48] This other tradition does not overtly challenge or subvert the Argentine myths of heritage, and the two might be able to coexist because they intersect in their focus on land and peaceful agricultural labor.

Lo criollo (Creoleness) captured the imagination of Argentine nationalists with its triumvirate of the gaucho, his bravery, and the pampa, all of which were disappearing under the new industrial order. Sarlo and Altamirano point out that the revalorization of "the Spanish heritage" found itself enveloped in the new myth of race, which led the way to new racialized discourses about the non-criollos, immigrants, or *yanquis*.[49] Territorialism was connected to the literal and figurative ground of the superior, spiritual race of the criollo, who was conceived in strict opposition to the immigrant. By this point in Argentine history, the gaucho had been trans-formed from the lawless roamer most nineteenth-century liberals perceived him to be into the quintessentially autochthonous figure. Thus, when the gaucho way of life was in decline, their culture was conflated to serve the mythmakers of the populists and national-ists. I should note here that the word "gaucho" itself is, to this day, ambiguous, as not only the origin of the term but also to whom exactly it refers is still a matter of debate. As Nicholas Shumway points out, in its narrowest sense "gaucho" designates the nomadic, often outlaw inhabitants of the great plains of Argentina, Uruguay, and Brazil.[50] In its broadest sense, and according to current usage, gaucho refers to "campesinos," or the rural working class. Gauchos played a contradictory role in the discourses and politics of the elite. They attacked and then incorporated into national culture, according to the interests of upper-class conservatives. The gaucho was first made the object of the civilizing mission and then depicted as the agent of that project, or at least as a mediator between civi-lization and barbarism. Urban intellectuals defined themselves against the people of the interior, the "barbarians" of Argentina. Yet they also wished to define themselves as traditionally Argentine in the sense that they had a national culture that was both distinct from the rest of Latin America and Spain and predated the wave of immigrants to Argentina.

Through the figure of the gaucho and his ambivalent position in Argentina, Gerchunoff tries to establish a place for the Jew in the cultural nationalist project. Gerchunoff himself adopts this position of expert and teacher, seen as important roles within the patriotic enterprise. The "gaucho *judío*" allows Gerchunoff to map out a sub-ject position that is at once separate from natives, Spaniards, other

Latin Americans, and other recent immigrants, revealing a potentially Argentine subject. The gaucho in *Los gauchos judíos* can be seen as a mediator, or translator, bringing verse and song to the frontier areas and making them acceptable for the cultured, more genteel Argentine criollo. In Gerchunoff, the Jew is linked to the very core of the gaucho's civilizing powers: his role as a *payador* (a sort of troubadour) informs newcomers about Argentine way of life.

While for the Centennial intellectuals criollismo was a vehicle of self-legitimization and the rejection of the foreign, for the popular classes, now displaced into urban areas, the rural utopia constituted an avenue of nostalgia or even antiurban rebellion. Kandiyoti does not develop the contrast between country and city in a detailed way, but it seems to me that the Centennial complaints against urban cosmopolitanisms are an important strategy for Gerchunoff to claim a space in the national imaginary for his immigrants. These "gauchos *judíos*," unlike the vast majority of Argentines and immigrants, know that true nationalism can only be learned through direct contact with the earth and that the city is only a place of materialist decadence.

In the story "The Anthem," misunderstandings between the Jewish newcomers and gauchos of Entre Ríos are examined in more detail. The Jews had greatly admired but misinterpreted most of the gaucho tales. Although they have adopted local dress, they are totally ignorant of their new country's political organization, mainly because of their lack of Spanish language skills. Trying to communicate with each other, the one word they all understand is "*libertad*," or liberty. One can hardly blame the immigrants for their lack of knowledge of Argentina, Gerchunoff suggests. Rather than there being some inherent racial problem or disinterest, it is simply a problem of communication, which will be naturally resolved in future generations.

Several stories examine Argentina as a possible promised land. Sometimes Gerchunoff presents an extremely optimistic picture, such as when immigrants arrive by train, and their drained and miserable appearances contrast sharply with their eyes bright with hope, while they are singing, "To Palestine/to the Argentine, We'll go to sow; to live as friends and brothers; to be free!"[51] However, in "Threshing Wheat," workers discuss whether they should have stayed in Russia after all, since so many children are becoming gauchos. Here, however, they have land, wheat, and livestock and can live in peace. Despite the increasing loss of religious tradition, then, Argentina still seems to be the best alternative. Unfortunately, in the following story, a cloud of insects attacks and destroys the entire orchard, leaving the village feeling desperate. Gerchunoff does not draw any definite conclusions about Argentina as a final Palestine; rather, the country seems to be the best option available, despite the difficulties.

The traditional happy ending in fiction is marriage. However, often marriage is not the final culminating act since the intricacies of the love stories in *Los gauchos judíos* do not end happily; that is, the conflict between family and lover is not resolved. In "The Sad and Lonely One," the lovely Jeved, reminiscent of women in the Bible, prefers to stay alone. Despite her many male admirers. She suffers from nostalgia for her childhood years in a picturesque city on the Black Sea, where she never had to do any farm work, and she is unable to forget her first love there. All of a sudden, the flute of a crippled boy, Lázaro, brings her back to the present. Deeply moved by his music, she finally forgets about her former sweetheart and eagerly starts awaiting her new gaucho friend. Although they come from vastly different cultures, and despite her beauty and his deformity, the art of the music connects them and helps Jeved to adapt to her new land. In this sense, both Lázaro and Favel are mediators, translators between different civilizations.

The alternatives to traditional Jewish marriage offered by the narrators serve to break open the small private sphere of the domestic. Once the marriage does not assume the focal point of the story, the unifying and nostalgic concept of home is problemized.

In "The Siesta," Jacobo is caring for his pony, despite the Sabbath and Doña Raquel's warnings. Reb Zacarías passes by and begins to gossip about the latest scandal: a girl has eloped with a gaucho. Jacobo defends the girl's decision, praises the boyfriend, and softly sings a gaucho song. Rather than describing the girl's acts as a betrayal of her community, Gerchunoff presents us this tale as something unavoidable: the girl was lost a long time ago, since she did not keep kosher. Nobody in the village seems to seek vengeance or try to force the girl back to the Jewish community. Again, it is art and tradition, the boyfriend's horsemanship and songs, that make the Jews meekly accept the loss of the girl.

"The Song of Songs" is another love story, and in it, Jaime asks Ester's father for the girl's hand in marriage. We do not know if this match will be approved; Ester could apparently also marry a richer man. But this love seems to sprout from the earth: he gives her wild partridge eggs, they discuss the cornfields and the land rather than personal matters, and there seems to be a genuine physical attraction between the two. Maybe then, the story suggests, miscegenation between Jews and gauchos is not so terrible, as long as the love is pure and uncontaminated by material considerations.

A similar story is "The Story of Miriam," in which another interracial couple, Rogelio and Miriam, the daughter of his boss, fall in love, even though they can only understand each other in song. Her father has to dismiss him to protect the girl's honor, but one day the villagers see him galloping through the streets on a horse with Miriam seated behind them. The girl here clearly does not have her father's consent, but their flight is described as noble, with Rogelio proud and riding high and Miriam staring defiantly at the people with burning eyes, while the sun colors the dust golden. In spite of its title, "Lamentations," this story is also deeply ambiguous. On the one hand, the colony is recalling the loss of Jerusalem, but at the same time two children, too young for prayer, kiss for the first time. The tale of displacement and loss is thus retold as a story of death and new beginnings, as cyclical and unavoidable as nature itself.

For the reader, the anticipated happy ending to the immigrant journey is a comforting idea at the end of a sequence of events that may tell of persecution and suffering, but it is still pervaded with the promise of a final reward of stability, marriage, and a new identity as an Argentine. With Gerchunoff, however, discourses surrounding assimilation suggest that the process undertaken by most immigrants implies the erasure of their "original," "authentic" identity, to be replaced by a new one. In order to enter into this new bond of symbolic kinship, one must successfully let go of one's original blood ties. The Jewish girls flee with gauchos out of love, not to escape hunger and poverty. They choose to become something new, to "translate themselves" into Argentine girls, and in order to do so they have to abandon their local culture and families. *Tradutore tradittore*, as the saying goes.

A translation is a text written in a well-known language, which refers to and represents a text in a language, which is not as well known and is never a simple technical (linguistic) transfer of systems; it is always a process invested with complex political, economic, and personal interests. Not coincidentally, the fifteenth and sixteenth stories in the collection, in which betrayal and the cutting of original blood ties are central, are the most violent in the collection. These are not stories of woman's love and betrayal but of the male dilemma regarding conflicting traditions. "The Herdsman" can be read as a reverse story of assimilation in which the gauchos, not the Jews, get into trouble because of the Argentine law's negative attitude toward gaucho values of loyalty and bravery. In other words, Gerchunoff here breaks open the assumption that gaucho equals Argentine identity; the urban Argentine as exemplified in the Argentine law cannot understand the legal code of the rural gaucho. The Jewish colonists, however, sharing the land and living with them, are able to understand these values. Don Remigio Calamaco, an old gaucho, nostalgically remembers his former days as an excellent warrior and is highly regarded by the Jews for his tales. Don Remigio's son loses a horse race and gets into a fight, which he also loses. Rather than see his son surrender, Don Remigio prefers to kill the boy himself, plunging his own dagger into the boy's neck. The Jewish villagers admire Don Remigio since they have learned that bravery is the primary gaucho trait; bravery was the race's chief source of nobility and its true poetry. The police, however, think otherwise and Don Remigio ends his days in a prison cell. The Jews understand the gaucho's dilemma between traditional lifestyle and submission to the nation's laws, but fortunately, they are not obliged to choose sides. It is Argentine law enforcement that decides Don Remigio's fate and thus betrays its own traditions of gaucho honor.

"The Death of Reb Saul" is based on the death of Gerchunoff's own father in a quarrel, an event that caused his mother to abandon the colonies and move to Buenos Aires.[52] In the story, Reb Saul must first pray and then saddle his horse for plowing. His servant, the gaucho Don Goyo, takes offense at his master's insistence that he use a certain horse and draws his knife into the man's chest. Don Goyo walks away, but later Reb Saul's corpse is discovered, resembling the figure of one of the great prophets. The story glosses over class differences—Don Goyo is Reb Saul's servant—as well as cultural differences. The reader is informed that Reb Saul barely spoke any Spanish and could possibly have misunderstood the gaucho's anger. The violence seems less important than the aesthetic: the women's lamentations, the victim's long hair, full beard, and white tunic.

Like Don Remigio's son, a Jewish immigrant too can become a victim of gaucho pride, but here, however, there is no mention whether Don Goyo is ever apprehended for his crime. Miscommunication is again central and eventually irresolvable. The gaucho cannot comprehend the importance of prayer, and Reb Saul cannot comprehend Don Goyo's sensitive attitude toward horses. Fortunately, the Jews do not have to choose between condemning Don Goyo, thus rebuking gaucho traditions, accepting what happened, and thus justifying the patriarch's murder.

Probably the best-known story, and the final one I will discuss, is "Camacho's Wedding," loosely based on an episode in *Don Quixote*. In the introduction, the rabbis mention Spain as a possible destination for their immigration, but they eventually decide to go to Argentina. In this story, Gerchunoff's love for Spanish Golden Age literature is most apparent as he appropriates certain elements of high Spanish culture for his own purposes.

In yet another tale of a twisted love affair, the story makes the reader understand that the handsome Gabriel from San Gregorio is a far more attractive and suitable husband for the lovely Raquel than her rich but unattractive suitor Pascual Liske. The couple performs a traditional Jewish dance, and the rabbi reads the marriage contract that conforms entirely with the sacred laws of Israel. Instead of dancing with her clumsy husband, Raquel dances with Gabriel, and a little boy, the most "gaucho" of all settlers, tells her that Gabriel is planning to elope with her. Many hours later, the boy informs the wedding guests that he has seen the runaways in a buggy, driving away fast. Liske accuses the bride of adultery, but the Shochet (ritual slaughterer) points out that she was not even married for one day and that therefore she can be divorced according to Jewish law. In the final paragraph of the story, Gerchunoff directs himself to the reader in an imitation of Cervantes' style. Fierce and arrogant gauchos and wife stealers, as well as rabbinical scholars, can all coexist. He asks the reader to remember his name, just as Cervantes remembered the name of Cide Hamete Benengeli and gave him all due credit for the original Camacho story.

Gerchunoff's fascination with the literature of Golden Age Spain, particularly Cervantes, is well known. He was but one of many intellectuals of the Centennial who turned to Spain in reformulating *Argentinidad*. Rather than trace the specific intellectual influences that might have led Gerchunoff to his notions of Hispanic culture, I would like to consider what is at stake in this conception. Raquel is forced to accept the rich man's marriage offer. Although she is Jewish and lives in the Jewish colony, the wedding, traditional though it may be, is already a place of cultural negotiation, where a mix of Russian, Jewish, and Argentine songs can be heard. More important, the match itself does not seem to be consistent with a true religious spirit. The bride's family insists on the match for financial reasons. Liske's money is inherited, not achieved through any special merit of his own. The family is ostentatious, preparing an ornate feast and mentioning the price of everything. Nothing is good enough for their only son—the word is given in Hebrew as "*benijujid*"—while the obvious unhappiness of the bride is disregarded. Interestingly enough, it is neither Argentine law nor the corrupted families that are able to offer a solution; rather, it is the shochet who understands Liske's unattractiveness and insists on a divorce according to Jewish law. Again, the "gauchos *judíos*" are not united in a supposed rejection of Argentine law, under which divorce would not be permitted. They once again negotiate their character: Jewish rules allow the girl to assimilate and become an Argentine. The establishment of a unified national community is frequently carried out by the delineation of a national character, and in Argentina, the Centennial generation looked back to the gaucho to emblematize this figure.

The cultural nationalists of the early twentieth century recast the gaucho as a civilizing force, a way of opposing the immigrant class. The positions of immigrants and gauchos had changed to such an extent since the days of Sarmiento and Alberdi's call for immigration that rather than the immigrants being viewed as the element of civilization that would tame "barbarian" gaucho culture, the gaucho was now invoked as the true Argentine who would keep the barbarous immigrants at bay. Turning the gaucho into a revered icon was part of the intellectuals' search for a historically rooted culture that would contravene rising mercantilism with more enduring values and counter the immigrant onslaught with a strong "autochthonous" tradition. The national character is a self-propagating fiction, however, and it is a fiction with very real functions, as this icon can be used to cultivate a specific sense of community.

However, one should be careful not to insist on a complete separation between "national" and "foreign," between Argentine "gaucho" and immigrant as if they were two isolated, self-contained entities. Although there certainly were plenty of anti-immigrant actions taken, one should keep in mind that the phenomenon of immigration itself was never under attack. Rather than a struggle between two separate entities, I would characterize the debates between cultural nationalists and immigrants as "negotiating" the terms of a new Argentine identity in the sense that Jeffrey Lesser uses the term in his study.[53] By using the term "negotiating," it should by no means be inferred that this process took place in a neutral place among equals. We are faced with the limits of performative identities when we consider that not only are immigrants potentially limited in the identities that they can inhabit by knowledge of only one particular linguistic register (a "lesser" register such as Yiddish or colloquial Italian), but they also are limited by nonstandard grammar and accented, "incorrect" Spanish. Not speaking "well" predisposed the immigrant to a position on the margins of citizenship. The registers of the minor tongue were pathologized and criminalized by the "scientific" discourses on language, reinforcing the pervasive identification of grammar, law, and the state. Being a criminal was thus a way of remaining an alien within the space of the national language. Nationalist writers like Lugones and Rojas reinvented the gaucho myths and traditions, while immigrants actively participated in inventing and negotiating new national identities.

Summarizing, one can say that the elites of the Centennial were concerned with a culture in formation. But rather than claiming that they were the sole keepers of "Argentine traditions," they were more prone to speak about refunctionalizing and reinventing the nation. History is thus transformed into a laboratory for future national morality, a patriotic lesson that would welcome and educate new audiences. One could even argue that, within this project, everyone can learn this nationalist discipline. It is also important to keep in mind that in some instances, cultural nationalists and immigrant groups worked together. For instance, as Sergio Pujol points out, Jewish publications such as *Vida literaria, Babel, Cuadernos del oriente y occidente, Cartel,* and *Israel* also published non-Jewish writers, forming the avant-garde in Argentine cultural journalism.[54] Even so, something of their language and old country ways remained. A purist like Ricardo Rojas occasionally collaborated in *Vida nuestra,* a Jewish magazine, while Russian-Jewish Alberto Gerchunoff's friendships with conservative cultural nationalists are well documented.[55]

Gerchunoff's insistence on the use of a classic Castilian, as well as multiple references to Cervantes, shows an insistence on demonstrating an authentic relationship to high culture as part of Argentina's heritage. His positioning of the Jew as a part of every Spaniard is interesting, predating a time when this notion became a common topic of discussion among writers such as Américo Castro. By turning the Camacho wedding plot, clearly associated with Golden Age Spain, into a forbidden love affair between a Jewish girl and gaucho boy, Gerchunoff points to a historical contact that came to a close, officially at least, in the early 1600s. If classical Spain already was in close contact with the Jewish Other, how then could Argentina feel threatened by these same people and consider them outsiders? Both Jews and Argentines share the same history and origin in Spain, and both came to break the ties with that country in order to look for a better life. The cohabitation of Jews and Christians in Spain, known as "convivencia," is the Spanish metaphor for cultural hybridization. If Argentines already were a mixed race before the influx of immigrants at the turn of the century, then the new immigrants, whether European or not, could not possibly taint the national stock. Never mind that in reality Gerchunoff's "gauchos *judíos*" came from Eastern Europe; in this work, they still claim a Spanish and Sephardic heritage.

This might explain the total absence of Yiddish words in *Los gauchos judíos*, despite its regionalist impetus. The persistent avoidance of any local or ethnic speech bridges "Jewish" and "Hispanic" heritages. Yiddish, the language without prestige of the lower classes, must be translated to Spanish, the "right" linguistic code, and thus the "Spanish" language became the portable homeland of the Jews.

I suspect that Gerchunoff's insistence on a "pure" Castilian is also meant as a vindication of the immigrant, whom the cultural nationalists accused of distorting and destroying the Spanish language. As previously discussed, the emphasis on a pure, national "mother tongue" in criollo Argentine discourses was part of the reaction against the immigrant influx. Correct or traditional criollo language is central in both the regionalist and social realist fictions of the day. Gerchunoff's language removes him from the Yiddish-speaking community even as it implies intimate knowledge of that community. His narrative voice places him outside of the immigrant community, allying with the unmarked and linguistically unaccented world of his metropolitan readers. In addition to the immigrants' Argentinization, Gerchunoff also narrates his own transformation into an Argentine and, more particularly, an Argentine man of letters. He resists identification as a representative of an ethnic community but claims a mainstream voice and identity even as he represents the immigrants and their world in his work.

In a move that references both presence and absence, the Jewish immigrants not only are made visible with the portrayal of particular customs and productive labor, but they are also conveniently absent, because their difference is unobtrusive and in many ways continuous with prevalent Argentine definitions of belonging: they do not live in the city, conspiring with dangerous mobs, but in a remote area, where they keep Argentine traditions alive.

An idyllic pastoral setting is not just a nostalgic way of remembering; it is also a means of criticizing the metropolis and its embrace of capitalist, "unpatriotic" ideology. Novels create the countryside as an answer to the fragmented and dangerously cosmopolitan character of the new metropolis. Previously considered backward and barbarous, the countryside now becomes the heart of "true" national identity. The mapping of "progress" depends on systematically inventing images of archaic time to identify what is historically new about national progress.[56] In other words, the nostalgic musings for the gaucho past were preconditioned by the turn-of-the-century mass immigration.

Gerchunoff's immigrants are nestled deep in this true Argentina. His carefully polished Spanish and complete omission of his native Yiddish, his use of Cervantes as a legitimizing subtext, and his use of the positive immigrant suggest that he was trying to create a space for his Yiddish immigrants within Argentine Centennial thought rather than pretending to write a *costumbrista* work describing their peculiarities.[57]

CHAPTER 4

BRAZIL AND ITS DISCONTENTS

ROMERO AND TORRES

THE HISTORIOGRAPHY ON EUROPEAN MIGRATION TO BRAZIL IS RICH in references to the official policies adopted by this country since the nineteenth century. However, few comprehensive studies have analyzed the evolution of pre–World War II immigration ideologies, expressed through legislation, official institutions created to manage the flow of immigrants or the writing of established public figures, as well as counterresponses by immigrants themselves.

As described in my first chapter, in the late nineteenth century, Brazilian supporters of immigration challenged the traditional view that race and climate combined to produce degraded and backward nations. A Brazilian thesis of white acclimatization in the tropics emerged, running counter to the common European view that for climatic reasons the white race was unable to work and thrive in extreme heat and that the low productivity and birth rate of the Brazilians was due to permanent features of climate and race.[1] While some intellectuals defended immigration ideologically, other more nationalistic views emerged in the early twentieth century.

My discussion will start with Sílvio Romero's 1906 attack on "the German danger," *O allemanismo no sul do Brasil, seus perigos e méios de os conjurar*, followed by a brief discussion of Alberto Torres's main arguments in which he denies the importance of "scientific race" or, rather, substitutes for that category a "national character." Graça Aranha's novel *Canaã*, published in 1902, plays with these fears of German separatism, racial degeneration, and national character and tries to solve this debate by referring to a chronologic and geographically mythical place: a promised land, Canaan.

I will then conclude this chapter by examining the "cosmopolitan" way in which immigrants were imagined in Brazilian "urban writing" of the 1920s, focusing on *Modernismo* in São Paulo. I suggest that although the *Modernistas* were certainly deeply involved in new, shocking ways to express their reality, they also form part of a larger debate concerning which "national" and "foreign" cultures are "suitable" to import and export.

THE GERMAN DANGER

As Jeffrey Lesser observes, to ask questions about the public construction of immigrant ethnicity opens windows to Brazilian national identity.[2] Acculturation of immigrant groups has not been recognized by both outsiders and insiders until very recently. Generally speaking, most studies of Italian or other ethnic groups concentrate on the uniqueness of their particular experience and rarely on policies and racial or ethnic ideas of the host society. Lesser shows that racial prejudice existed side by side with the acceptance of undesirable groups for economic reasons and that these elements provide an important model for the study of ethnicity in a society such as Brazil's, which officially denies its multiculturalism. Already in 1888, in the heyday of positivist thought, Sílvio Romero used the language of chemistry in asserting that immigration was a social "reagent" to be handled with the greatest of care since Brazil had "a singular ethnic composition."[3] All immigrants and their subsequent generations, no matter their origin, argues Giralda Seyferth, faced the tension between becoming Brazilian and retaining an original ethnic or national identity.[4] This tension has been expressed in a variety of forms. Social clubs, schools, and specialized commercial sectors associated with specific immigrant groups remained strong. According to Seyferth, between 1886 and 1936 around 280,000 native German speakers entered Brazil. She points out that European immigrants were expected not only to provide the precious labor for agricultural and artisanal production but also to assimilate into Brazilian culture and whiten its racial pedigree. For these immigrants and their descendents, the notion of "Deutschtum" or "Germanness" did not conflict with the idea of being Brazilian since they considered their adoptive country to be an ethnically plural state rather than a unified nation. For Brazilian nationalists, on the other hand, this idea of *Deutschtum*, of a Pan-German identity, began to seem dangerous, threatening the unity of Brazil. Thus, despite their whiteness, since Germans were suspected of being clan-oriented and unwilling to assimilate, non-Catholic, and non-Iberian or "Latin," they became the target of nativist backlash in the early twentieth century.

Romero's essay opens with an ironic dedication to Baron Rio Branco and Joaquim Nabuco. Romero is explicit about the reason he has chosen to dedicate his work to these two politicians: "Os dois modernos estadistas Brasileiros, que poderão, se o quizerem, iniciar a politica de reacção contra o péssimo systema seguido até hoje da colonisação alleman do Sul do Brasil" (The two modern Brazilian statesmen who could, if they wanted, start a policy against the very bad system followed until today of the German colonizations in the South of Brazil). His second dedication to the newspaper that publishes his is more sincere: "Á patriótica redacção d' *O Jornal do Commércio* do Rio de Janeiro. A folha brasileira que mais serviços tem prestado na questão do perigoso allemanismo do Sul do Brasil e os poderá prestar ainda maiores"[5] (The patriotic editorial board of the Jornal do Commércio of Rio de Janeiro, who have most given their help in the question of the dangerous "Germanness" in the South of Brazil and can provide us with even more help). Romero then continues to say that there are only two ways to create strong, prosperous, and cultured nationalities: either let the current nations follow their own course until the causes of their backwardness are removed, or else eliminate those causes—in other words, eradicate the existing populations.[6] At least, Romero is of the opinion that it is best to choose the first option, since eradication of the black and indigenous populations in Brazil did not seem feasible. He then contrasts two forms of colonization: the Brazilian and the Japanese. The former is anachronistic, causes inequality between the North and the South, and is dangerous because it stimulates regionalism. The latter, in contrast, is the one that he feels should be followed. It does not stimulate racial mixing, and if Japan is taken as the example, Brazil could become a cultured modern nation with high moral qualities and eventually, a world power.[7]

According to Romero, Brazil will also urgently need to change its educational system, focusing only on positive national characteristics. In addition, he claims, it should follow the policies of the United States in which all foreigners have to use the national language and, if they want to be assimilated, must be white. Previous Brazilian governments had done nothing of the sort since 1825, when the first German colonies in the south were founded. These colonies, it was thought, will in the future cause a serious threat to the country's existence; by allowing the German language and culture to monopolize the area, the chance of the colonizers returning to their country of origin remained a real possibility: "Não canso de repetir: tal systema póde ser óptimo, e o é, por certo, do *ponto de vista allemão*; mas é péssimo, perniciosissimo, do *ponto de vista brasileiro*"[8] (I will not tire of saying it: such a system may be optimal, and it is, certainly, from the *German point of view* but it is terrible, extremely pernicious, from the *Brazilian point of view*).

Romero fears that the German race has a tendency to isolate and eventually separate itself and remembers the decadence and fall of the Roman Empire.[9] How then, he asks, can Brazil be so sure that they will not separate themselves again?

> Metteram-se por todas as provincias, como hoje se mettem pelo nosso Brasil meridional . . . Desde então, os dias de Roma estavam contados, e os vencedores, os destruidores, os herdeiros do imperio só não eram conhecidos dos cegos optimistas, dos patrioteiros de vistas curtas, que não falham nunca entre os povos que vão morrer.[10]

> [They flocked to the provinces, as today they flock to our South . . . From then on, Rome's days were numbered, and the victors, the destroyers, the heirs of the empire were unknown only to the hopeless optimists, to the short-sighted patriots, who never lack among those peoples on the verge of death.]

In the United States, German immigrants cease to be Germans, and they accept their new nationality. While the elderly still speak German, their children are born Americans, and after one or two generations, they forget the old language. Only in Brazil, in the German colonies in the South, do the immigrants never bother to learn Portuguese. According to Romero, the people in those colonies are Brazil's adversaries since they refuse to change their language and customs, implying that they do not respect them. These immigrants care less about Brazil than about a war between China and Japan. They refuse to participate in the local politics of Santa Catarina and Paraná, as if they were foreigners, as if they were members of African tribes.[11] Altogether, he warns his readers, there are about 380,000 people of Germanic origin in Brazil:

> Ora, os allemães do Brasil são, *socialmente*, completamente distinctos e independentes dos nacionaes. Teem outra lingua, outra religião, outros costumes, outros habitos, outras tradições, outros anhelos, outros generos e systemas de trabalho, outros idéaes. . . . Logo, estão presos a nós sómente pelo *laço do terrritorio*; porque mesmo de um *laço politico effectivo* não se pode falar desde que se sabe que elles não tomam a mínima parte em nossa vida por esse lado.[12]

> [Now, the Germans in Brazil are, socially, completely distinct and independent of local Brazilians. They have another language, another religion, other customs, other habits, other traditions, other hopes, other forms and systems of labor, other ideas . . . Then, they are our captives only through *territorial bonds*; because we cannot talk about effective *political bonds* since, as we all know, they take no part in our life in that respect.]

It is the lack of desire for miscegenation that makes the German immigrant so dangerous in Romero's view. For this reason, the Brazilian government should force them to integrate by forbidding the use of German and forcibly relocating the population. It is remarkable that Romero converts the "pure" Germans into an undesirable race precisely because they are "pure" and therefore unpatriotic. Apparently, to be of mixed origins (which origins, evidently, remains unspecified) has ceased to be shameful, and the lack of purity is precisely what makes the subject a pure Brazilian. Rather than hopelessly trying to achieve European "pureness," Romero suggests an alternative racial model to imitate: that of the United States.

ALBERTO TORRES AND THE BRAZILIAN NATIONAL PROBLEM

Another fierce opponent of immigration was Alberto Torres (1865–1917).[13] His xenophobia was legendary, and he even drafted his own new constitution to oppose the 1891 version, with its U.S.-style federalism.[14] Although *O problema nacional brasileiro* was only published in 1914, Torres's ideas had circulated widely in journals for the previous two decades. He suggested the idea of an "inverted Babel," an increasing homogenization and growing nationalism, to define a "Brazilian" character. He defended the Brazilian worker, who was constantly marginalized in favor of European immigrants. Most relevant to the topic of immigration is a study originally published in 1912 titled "Canaã." A self-proclaimed nationalist, Torres worried about the "disorganization" of Brazil, which he described as not one people with one conscience, but always exploited by foreign forces:

> A necessidade de capitais e de braças estrangeiros era um dos abrigos a que se tinham acolhido a nossa indolência . . . em face dos problemas da nossa economia . . . Esse apelo não tem por si a apoio de nenhuma teoria . . . sem capacidade para dar soluções práticas, os polticos Brasileiros continuam a comprometer os povos nos riscos de suas concepções fantásticas.[15]

> [The need for capital and for foreign labor was one of the layers of protection that our indolence clung to . . . in the face of the problems of our economy . . . This call itself had no support from any theory . . . without the ability to provide practical solutions. Brazilian politicians continue to expose the people to the risks of their fantastic concepts.]

As seen previously, Torres denies any supposed degeneration of Brazilian citizens. What makes the country so vulnerable is the fact that foreigners are welcomed with open arms so that they can exploit and destroy the land, and subsequently, "abandonada a terra, e não cuidando da nação, abandonamos a Pátria, porque a Pátria é a terra, como habitat, mas principalmente, para o sentimento e para a razão, a nação, isto é, a gente"[16] (the land abandoned, and not caring for the nation, we abandon the Patria, because Patria is the land, a habitat, but principally, for feeling and for reason, a nation—that is, a people). He dismisses race as a factor in the creation of a culture: "Aprender com alemães, com americanos, com franceses, com ingleses, e com brasileiros, quando for possivel a ser Brasileiros eis a fórmula ideal do nosso cosmopolitismo mental"[17] (To learn with Germans, with Americans, with the French, with the English, and with Brazilians, when it was possible to be Brazilian, see here the ideal formula of our mental cosmopolitism). Torres accuses the government of being careless and of wanting to create an "intellectual, cultured Brazil" before the country has even stabilized its own territories. They plant imported seed without knowing how to harvest it; they import and cultivate foreign fruits, neglecting the native ones.[18] A strong nationalist policy will be the only way to lead Brazil out of its crisis: "Esta bela noção afetiva da Pátria, que mostra, nas migrações de selvagens e de bárbaros, como um astro orientador, a terra ignorada e formosa, onde se oculta a promessa do reino de Javé . . . inspira ao aberrante símbolo do Canãa, para imagem do nosso ideal patriótico"[19] (This beautiful affective idea of Patria, which reveals, in the migrations of savages and barbarians, like the north star, the fertile and ignored land, where the promise of Yahweh's kingdom hides . . . inspires the aberrant symbol of Canaan, for the image of our patriotic ideal). This promised land will have to be of mixed racial origins, but Torres sees no problem here, since, according to him, mixed populations are more vital, live longer, and are more fertile than the so-called superior races.[20] The immigrant communities Torres dislikes the most are the German colonies of the south, which, in spite of their wealth, are artificial because they keep their German traditions alive: "Do colono alemão nada é preciso dizer. Ele so recomenda pelos próprios o merecidos títulos . . . o caso do alemão brasileiro é mais uma prova da falência da doutrina da superioridade das raças"[21] (Concerning the German settler there is nothing to say. He only associates with his own or with those with proper titles . . . the case of the German-Brazilian is yet another example of the failure of the doctrine of racial superiority).

While Romero believes in different races and racial characteristics, he does not believe in a simple substitution of natives for Germans. For him, German blood alone will not suffice to modernize Brazil. Torres claims that the national subject is being neglected

and carelessly replaced by a dangerous foreign type, whose civilizing effects remain to be seen, since it is unclear how exactly the
Germans will mix with the native population. They both agree that
any successful immigration policy will have to take into consideration the "true essence" of the country. As I will show, this is one
of the main preoccupations of Graça Aranha's novel *Canaã* (1902),
one of the first works that overtly engages this search for a perfect
race through the process of *mestiçagem*.

GRAÇA ARANHA'S CANAÃ

Graça Aranha's elite credentials were much more established than
those of Torres or Romero.[22] In the 1880s, he attended the University of Recife, where he came into contact with the latest European
books and ideas, including those of Spencer and Darwin, accompanied by their racist overtones. He served for several years as a municipal judge in Porto do Cachoeiro, in Espírito Santo—the city that
was later to serve as the setting for his novel. Porto do Cachoeiro
was a center of the expanding coffee industry and was dominated
by a German immigrant population. In the 1890s, Graça Aranha
entered the diplomatic corps and served with two of Brazil's leading
international figures: Joaquín Nabuco and Baron Rio Branco. As
part of his service in Brazil's foreign ministry, he traveled to both
Rome and London. By this time, urbanization was already quite
advanced, and the memory of slavery (at least for the white elites)
seemed more distant than in Aluísio de Azevedo's time. Thus, the
Brazilian countryside of Espírito Santo is reinscribed as the utopian
space where true national harmony can be found.

Drawing on biblical frames (the title is characteristic of the
heavy-handed tone of the work) and continuing the model of the
nineteenth-century romance novel, racial miscegenation is proposed as the inevitable and ultimate solution to Brazil's lack of
national unity. Despite the corruption, cruelty, and racial inferiority of Brazil, progress is seen as unstoppable, and the creation of
a new national race seems imminent. As I mentioned earlier, the
novel is situated in a small city in the interior of the state of Espírito
Santo, in a zone predominantly colonized by German settlers. The
town of Porto do Cachoeiro is the crossroads of the two races and
thus symbolic of the larger transformation taking place in Southern

Brazil. *Canaã* is not a regionalist novel, interested in describing the typical or the picturesque, but rather a novel of ideas on the possible racial future of Brazil. The hero of the story is Milkau, a German immigrant from Heidelberg, who, in spite of his obviously cultured background, decides he would rather work the land than dedicate himself to commerce in the city. This perfect immigrant is optimistic and dreams about a new tropical race, a fusion of all people. He longs for Canaan, the promised land, where all men will live in harmony.

Milkau and his fellow countryman Lentz have long discussions about social and philosophical topics. Milkau deplores the destruction of Brazilian traditions and does not believe in the doctrine espoused by his friend, who thinks that to renew Brazil means to import superior races. The debate focuses on the significance of the country's racial heritage and debates its cultural future. Is Brazil truly inferior to Europe due to the African and American Indian elements in its society? Can Brazil develop autonomously, or must it rely on European culture for help and guidance? Can immigration strengthen society, or will the settlers isolate and separate themselves? Will immigrants supplant the native Brazilians and flourish, or will the Brazilians assimilate the Germans? The novel portrays the dilemma between the elite's desire to imitate European culture and capitalism and its desire to develop a viable national culture and identity. The more the Brazilian elite clings to European economics and European thought, the more dependent it becomes on Europe's economy and the more it will be forced to deny its own cultural and intellectual heritage. The ideological framework Brazilian intellectuals had imported from Europe faced its severest adaptive test when faced with the question of race.

All Europeans are not the same, Aranha insists. Milkau loves his new life as a coffee grower. His Tolstoyan ideas stand in stark contrast to those of Lenz, an aristocrat, whose Nietzschean philosophy and disdain for the inferior quality of Brazilian culture serve as a counterpart to Milkau. The two immigrants are exponents of entirely different cultures. One offers warlike exploits, massacres, and bloody sacrifices; the other, a simple farmer, offers fruits from the earth and flowers from his garden. Milkau seeks to generate love and unite the spirits. Lentz, however, sees life as a struggle, rife with criminality. All human pleasures, he tells Milkau, taste of blood; everything represents victory, the expansion of the warrior.[23] These two young men, the vanguard of European culture, voice the concerns of different sides of the racial question. Lentz represents the hard-line European stance: in favor of eliminating the "inferior" Brazilian race and populating the country with "superior" Germanic peoples. He claims, "Não é possível haver civilização neste país . . . A terra só por si, com esta violência, com esta exuberância, é um embaraço imenso"[24] (Civilization is impossible in this country . . . the land itself, with this violence, with this exuberance is a terrible impediment). The supposed degeneration of the Brazilian people is clearly equated with its impure, nonwhite blood. Bad genes foretell cultural decay. Lentz is proud of his race and paints a vivid portrait of a Brazil that will be biologically and culturally European. Milkau, in contrast, feels sympathy for the new country and worries that the city will crumble to ruins, surrounded as it is by foreign colonies that are choking it by degrees until, someday, they will ruthlessly conquer and transform it. He is surprised that the German settlers never learn any Portuguese and that all instruction is carried out in German.[25] Local Brazilians, in contrast, all manage to learn German.[26] Still, he insists on the inevitability of the transforming power of European culture, assuming that the Europeans should govern and lead:

> Mas isto é a lei da vida e o destino fatal deste País. Nós renovaremos a nação, nos espalharemos sobre ela e a engrandeceremos para a eternidade . . . Falando-lhe com a maior franqueza, a civilização dessa terra está na imigração de europeos; mas é preciso que cada um de nós traga a vontade de governar e dirigir.[27]

> [But this is natural law and the fatal destiny of this country. We will renew ourselves as a nation, we will spread and we will aggrandize it for eternity. Speaking frankly, the civilization of this land lies in European migration; but it is necessary that each of us bring the will to govern and rule.]

Milkau is a late exponent of the nineteenth-century father/hero, who seeks to domesticate the virgin land and father a new race. Aranha seems to share Sílvio Romero's fear of a "German Danger."

The German immigrants in *Canaã* dedicate themselves to improving their economic situation and are totally uninvolved in local politics. These local politicians serve as mouthpieces for different views concerning Brazil's dependency on Europe. The municipal judge Maciel is a good but weak character, unable to stop the shameless corruption of his employees:

> Os senhores falam em indepência mas eu não a vejo. O Brasil é e tem sido sempre colônia. O nosso regime não é livre: somos um povo protegido. É o debate diário da vida Brasileira . . . ser ou não ser uma nação . . . Que podemos fazer para resistir os lobos? Com a bondade ingênita da raça, a nativa franqueza, a descuidada inércia? . . . Temos o que merecemos.[28]

> [People speak of independence, but I don't see it. Brazil is and has always been a colony. Our regime is not free: we are a protected people. This is the daily debate in Brazilian life . . . to be or not to be a nation . . . What can we do to resist the wolves? With the innate goodness of our race, the native frankness, the indolent inertia? . . . We have what we deserve.]

The judge sees Brazil as a colony, doomed because of the childlike trust and enthusiasm of its inhabitants. A local landowner complains that the government does nothing for the Brazilians: everything is for the Germans.[29] A mulatto lawyer reacts strongly to the flood of immigrants and their culture and shouts that though the Brazilian elite may want to sell out the country, as long as there is a mulatto left, things will not run as smoothly as imagined.[30]

Clearly, the outlook for traditional culture and the mulatto is not optimistic, and the depiction of the local Brazilians is not at all flattering. At best, they are weak or lazy; at worst, they are ruthless exploiters of the immigrant communities. The mulatto is perceived by both the author and the novel's European characters as the predominant figure in the population, and notably, he is usually described in animalistic terms. Although Milkau feels sorry for the poor boy who serves as his guide the first time he arrives at Porto do Cachoeiro, this same child is also described as a "rebento fanado de uma raça que se ia extinguiendo" (withering shoot of a race in extinction), while another mulatto girl is portrayed dirty and lazy, "a própria indolência" (her own indo-lence). [31] Even the Brazilian landscape is "cansada . . . morria alí como uma bela mulher ainda moça"[32] (tired . . . dying there like a beautiful women still in her youth).

These people live in the past, unable to compete with the German colonies, remembering the good old days when they were still slaves and had been properly cared for by their owners. [33] These descriptions are merely manifestations of a fundamental assumption underlying the entire novel: the cultural and biological degeneracy of the Brazilian compared to the European.

Through the main character's reaction and observation of the people, the overly optimistic tone expressed in the beginning of the book gives way to basic greed and jealousy. The second part of the novel focuses on Maria Perutz, a German colonial servant girl. Milkau first meets her at a colonists' party at which Germans dance their own dances, much to the chagrin of Felicíssimo, a mulatto from Ceará, who dances in his own tradition but cannot find a partner, since none of the German girls know the steps.[34] It seems that the perfect couple would be Milkau and Maria. After becoming pregnant, Maria is thrown out of the house of her German boss. After many hardships, she is aided by Milkau. She gives birth in the forest, and her child is eaten by a horde of wild pigs. Unjustly condemned by public opinion as a murderess, she is imprisoned. Milkau believes in her innocence and helps her escape, and the novel ends with them fleeing together through the wilderness in search of the true Canaan. Maria's child is devoured by nature itself. Maria, the perfect *colona* and the hope for a racially integrated future, is cast out by her own people, and her baby is literally a European fruit devoured by wild animals in the forest.

It is in the context of the question of the national and the foreign that the central symbol of the novel emerges—that of Canaan, the promised land. Milkau is clearly the symbolic figure of Moses who will lead his people to the promised land. It seems that one has to let go of Europe and its faults and failings and resurrect its glories in this new land. Maria's flight finally turns Milkau away from the city, and like Moses, he brings her to the mountaintop where they envision a glimpse of Canaan. At the end of the novel, Milkau then suggests that they stop there to maybe see the true Canaan, to wait for it to come with the blood of redeemed generations.

The incipient nationalism of the Brazilian characters and the desire to Europeanize Brazil are conflicting tendencies that are never resolved in the novel. *Canaã* typifies much of the attitudes felt by Brazilian intellectuals toward German settlers, showing them as wealthy Europeans interested only in exploiting the riches and natural wealth of the land. Aranha also portrays the Brazilians in a negative light, describing the judges as corrupt and intent on draining the wealth of the industrious German farmers. The author takes a decidedly pro-European stance against the darker-skinned natives, whose mixed blood he feels possesses the more base human qualities. At the same time, however, he casts the Germans in a negative light when they shun one of their own due to an unwanted pregnancy. Aranha's novel thus attempts to deal with the differences between the two races and peoples by adopting European ideas and casting them in a Brazilian framework.

Most of the criticism on Graça Aranha describes his position as a mentor for Brazilian *modernistas*, two decades later, mainly because he organized, together with Paulo Prado, the Week of Modern Art in February of 1922. A good example of this can be found in a school edition of *Canaã*, in which the editor, Dirce Côtes Riedel, declares, "O imigrante quer esquecer os preconceitos da cultura importada, a pompa da civilização européia. Tal pensamento que em 1925 Graça desenvolve em *O Espírito Moderno* já está esbosado em 1901 e será uma tese esencial do modernismo brasileiro"[35] (The immigrant wants to forget the stereotypes of the imported culture, the pomp of European civilization. Such an idea that in 1925 Graça develops in "O Espíritu Moderno" [The Modern Spirit] was already outlined in 1901 and will be an essential thesis of Brazilian *modernismo*). Riedel suggests the novel represents an attempt to break away from European models and stereotypes a quarter of a century before *modernismo*.

Personally, I see very little connection between Graça Aranha and the *modernistas* beyond the advocacy of mixed cultures, partly European and partly indigenous. The *modernistas* would certainly approve of his promotion of a unique syncretism between races and cultures and the revalorization of a "national identity." However, as we have seen, those ideas were already circulating by the time of the novel's publication. Graça Aranha was by then an older and well-established academic whose support definitely helped the young *modernistas*. However, unlike a novel such as Mário de Andrade's *Macunaíma*, an open and polyphonic text, *Canaã* is extremely authoritarian, and the reader is never in doubt as to which side to choose. Much has been made about Aranha's involvement with the *modernista* generation of the 1920s, but unlike that generation's artistic output and ideals, *Canaã* is practically unreadable for the contemporary reader. Again, the only point of contact between the two, in my opinion, would be the celebration of *mestiçagem*—not any literary techniques. In that sense, *Canaã* can be regarded as a work of transition between nineteenth-century positivist thought and new nationalist ideas that would manifest themselves in the *Tropicalismo* of *modernistas* such as Mário de Andrade, Oswald de Andrade, or even Gilberto Freyre's *Casa-grande & Senzala*.[36]

Characteristic of Aranha's thought is the belief that every philosophy needs to be based in a truly Brazilian experience. He repeats this message over and over again, from *Canaã* through his article "O Espírito Moderno" ("The Modern Spirit"), published after the *Semana de Arte Moderna*, an event usually taken as the beginning of Brazilian *Modernismo*. In "O Espírito Moderno," he affirms that all culture came with the Portuguese founders but holds that this culture was quickly modified through the climate and racial miscegenation. It is this indefinite character that will enable the creation of a true future Brazilian nationality. European culture should not be used to transplant or to imitate Europe but as an instrument to create new things with elements that come from the earth, the people, and the savageness of Brazil.

THE CITY AND ITS DISCONTENTS: MODERNIST SÃO PAULO

As wealthy Paulistas started doubting the cultural superiority of many of these immigrants, they became more self-assured and began criticizing the supposedly inherent superiority of Europe. Intellectuals once again embarked on a search for a mystical "homeland" with one national race. By asserting that geography (i.e., nature) was the basis of race, "white" immigrants *in* Brazil would create a European-like national identity that would smother the native and

African populations with its superiority. As Jeffrey Lesser points out, even the word *raça* was fluid: it could refer to people (the human race) or animals (breeds) or, even more generally, species. The same word could describe a person's cultural identity or dehumanize him or her as a "half-breed." European countries had long been symbols of civilization when compared to Latin America's so-called barbarism and backwardness.[37]

This Brazilian nationalist impulse gained significant strength during World War I as the São Paulo region grew from an agrarian to an industrial economy based on coffee production. The transformation of the Brazilian intelligentsia was connected to this economic shift: new concrete conditions demanded a new attitude. The ideologies of the past could not describe these new developments. Although one cannot say that there is a clear break with positivist thought until the 1920s, we do see an increase in a more sociological attitude, as opposed to humanistic erudition.

"Whiteness" remained one important component for inclusion in the Brazilian "race," and as described previously, eugenics-influenced policy initially favored the entry of German, Portuguese, Spanish, and Italian workers as *braços para a lavoura* (agricultural labor). Yet, a fear of social and labor activism in the city, as well as concerns about the immigrants' desire to assimilate caused a modification in the language that described "Europeanness" as desirable. The impulse to make immigrants "white," regardless of their ostensible biological race, matched neatly with immigrant hopes to be included in the most desirable national category.

One result of the wave of immigration was the relatively sudden spurt of urbanism in Brazil. The population of the city of São Paulo, for instance, jumped from only 129,409 people in 1893 to 240,000 by 1900—a growth of nearly 100 percent in seven years. By 1907, Italians alone outnumbered Brazilians in the city by two to one. The federal capital to the northeast, Rio de Janeiro, was by this date a city nearing a population of 800,000.[38]

Fundamental to this phenomenal urban growth was the influx of foreign immigrants. Starting in the 1870s, travel expenses for immigrants were paid by the *fazendeiros* in the São Paulo region in exchange for a five-year contract after which they had to pay back their travel fees. Immigrants were ruthlessly exploited, and many fled to the cities. For instance, in the city of São Paulo in 1920, 35 percent of the inhabitants were foreign born, and in 1934 immigrants and their children represented 50 percent of the population of the entire São Paulo province.[39]

Edward Timms argues that the modern metropolis found itself in a precarious balance between coherence and disintegration where new alliances were constantly being formed and rearranged.[40] The only common language its inhabitants all shared was that of money. In the second half of this chapter, I will discuss some of these new urban alliances that were originally based on ethnicity but became increasingly more "Brazilian." I will start with *Fanfulla*, an Italian language newspaper, and then introduce a text that reflects the liberal elite's attempt to "domesticate" the foreign worker, *Brás, Bexiga e Barra Funda*, by António Alcântara Machado.

AN INVERTED BABEL

São Paulo, without any colonial tradition, and with its explosive growth, was occasionally seen as an inverted Babel. In this myth, the Old World is seen as decadent, fossilized, and oppressed, whereas the new world is mysterious, exciting, dynamic, and enchanting—the place where a new homogeneous race will be created. This new world is symbolized by the booming city of São Paulo, where people of all races coexist and interact.[41]

Unfortunately, in reality this New Babel was not harmonizing at all, and the vast majority of São Paulo's population lived in appalling conditions. Immigrants competed with former slaves and the so-called *caipiras*, people of mixed blood coming from the countryside looking to improve their conditions in the city. Some immigrant groups, especially the Italians, did manage to organize, through mutual help organizations, charities, and labor unions. Of all the destinations of Italian immigrants during the period of mass migration, no country captured the Italian imagination more than Brazil, and in particular the state of São Paulo. Italian perceptions of Brazil fluctuated dramatically over time inspiring lively debate among Italians in favor of and opposed to migration to Brazil. To some it was a vast land of the future, promising wealth and prosperity; to others, a backward wildness run by cruel *fazendeiros* who treated Italians as they did the slaves of the past. After alarming consular reports on the treatment of rural Italian laborers in São Paulo, the government of Italy enacted the Prinetti Decree, prohibiting the subsidized passage of Italians to São Paulo. From 1880 to 1930, approximately 1.5 million Italians immigrated to Brazil, primarily to the state of São Paulo, although significant numbers also settled in the southern states of Rio Grande do Sul, Santa Catarina, and Paraná. From 1888 to 1902, the height of Italian immigration to Brazil, approximately 942,463 Italians arrived in Brazil.[42] During these years, Brazil was the most popular destination for Italians, surpassing both the United States and Argentina. Coming at the tail end of European mass migration, the 1920s, both the governments of Brazil, in particular the state of São Paulo, and Italy reexamined their emigration and immigration policies. With the introduction in 1924 of a new restrictive quota system, the United States, which had, since the turn of the century surpassed both Argentina and Brazil as the most popular destination for Italian emigrants, was no longer an option, causing Italians to regard Brazil as one of the most attractive remaining destinations.

The Italian side of the debate on emigration was both a reflection of the perceived condition of Italians living in São Paulo, as well as a product of the changing Italian views toward emigration in general. It is striking how the "exotic" life on the coffee frontier captured Italian imaginations far more than the commercial-urban immigration experience, which, unlike the *fazenda*, was not unique to Brazil. It was this aspect that made Brazil different from the other popular emigration destinations. As David Aliano notes, there were constant references to the miserable lives of rural immigrants in the newspapers, whereas the equally grueling conditions in the city were rarely mentioned.[43]

Within Italy, the attitude toward immigration, formally perceived as a cause for national embarrassment, also changed. For instance, in one of his first addresses concerning the question of Italian emigration, Mussolini positively redefined the immigration issue as following: "Italian expansion in the world is a problem of life or death for the Italian race. I say expansion: expansion in every sense: moral, political, economic, demographic. I declare here that the Government intends to protect Italian emigration: it cannot be indifferent to those who travel beyond the Ocean, it cannot be indifferent because they are men, workers, and above all Italians. And wherever there is an Italian there is the tricolor, there is the Patria, there is the Government's defense of these Italians."[44] In this view, rather than losing vital manpower, emigrants, by retaining their Italian identity, expanded and strengthened the patria, creating a global empire. Interestingly enough, at the same time as this reorientation of policy in Italy, the state of São Paulo also began to redirect its policy toward immigration. In 1927, it eliminated its program of subsidized passage to São Paulo.[45] The most significant areas of disagreement between the Italian government's point of view and that of São Paulo involved the issue of Italian diplomatic intervention in labor negotiations among laborers and *fazendeiros* (wealthy landowners) within São Paulo, as well as the more recent Italian preoccupation of preserving and promoting Italian identity abroad. Assimilation, loyalty to Brazil, and an education in Portuguese were also stressed, in obvious contrast to the Italian government's desire to preserve *italianità* abroad.

Confronted with working days of fourteen to sixteen hours, this new proletariat quickly forgot its rural origins.[46] In this process of acculturation, immigrant groups tended to create a variety of institutions such as mutual aid societies, labor unions, social clubs, schools, churches, and newspapers, to help them cope in their new environment. One of the most important, yet most neglected, of these institutions was the foreign language press. These ethnic newspapers have received limited attention in the United States; in Latin America, they have been totally ignored. *Fanfulla*, the main newspaper for Italian immigrants in São Paulo, founded in 1894 by Vitalino Rotellini, was an important vehicle for the distribution of information within the Italian communities of Sao Paulo. Joseph Love has written extensively on the cultural climate and journalism of São Paulo in the 1920s and 1930s.[47] São Paulo had the greatest number of periodicals in the nation and was second only to the federal district in the number of people employed in journalism. In the decade of 1920 to 1929 alone, over five hundred journals and papers were founded. *Fanfulla* and the Brazilian paper *O estado de São Paulo* were the two largest newspapers of any kind in the state. *Fanfulla*, which at one time reached as many as thirty-five thousand readers, began as an eight-page morning daily and by 1913 had expanded to twelve pages.[48] In 1906, Vincenzo Giovannetti of the Buenos Aires publication *La patria* came to edit the paper. It continued publication daily until 1964, with the exception of five years during World War II. This liberal, anticlerical, and mildly antimonarchical daily sought to represent the interests of all social strata within the Italian community of São Paulo and Brazil.[49] A number of telling images recurred in the newspaper. Continual reference was made to Brazil's slave past, as the plight of Italian *colonos* was characterized as a new form of postabolition slavery. The contrasting image of Brazil, which also occurred frequently, was that of the "land of the future," which thanks to its natural resources and vast territory, always seemed potentially prosperous, forever a great nation in the making.

Much of *Fanfulla* was devoted to news about Italy, but increasingly, attention was given to the living and working conditions of Italians in Brazil, to the labor movement, and to the socialists. News from Buenos Aires was also prominent, reflecting the considerable impact of the Italian colony in Buenos Aires on that of São Paulo. There were also daily columns that focused on local news from different regions of Italy as well as sections of letters and commercial news. The advertisements clearly show a new urban aesthetic and frequently mix Portuguese and Italian. Many ads tout patented medicines for people who are too poor to see a doctor, while others show Italian steamship companies competing for the lowest fares. However, there are also ads for fine Italian wines, Brazilian beauty products, and modern objects such as sewing machines, and restaurants with both Italian and Brazilian cuisines.

Fanfulla served as a community bulletin board for Italian organizations. The São Paulo paper consistently favored the participation of immigrants in political life. Although it never explicitly recommended that the immigrants renounce their Italian citizenship, it did urge them to become Brazilian citizens. *Fanfulla* acknowledged the "prejudice against naturalization" among Italians and the "fear of being a traitor to Italianism." Nevertheless, it vigorously pursued its campaign for naturalization and political participation.

According to Sevcenko, the living and working conditions were so brutal that of the more than one million immigrants who came to São Paulo state between 1884 and 1914, almost half left the country looking for a way to improve their lot.[50] In 1922, for instance, *Fanfulla* declared it was against immigration, comparing the treatment of Italian newcomers to that of former black slaves. In that same year, the *Liga Defensiva Brasileira* (Brazilian Defense League) was founded by some two hundred coffee barons and quickly attacked the foreign press: "Aos indesejáveis e ao jornal italiano *Fanfulla* . . . tomarão severas medidas . . . se o mesmo continuar as repelentes infâmias de pasquineiro ignóbil contra nós e nossa terra"[51] (They will take severe measures against the undesirables and the Italian newspaper *Fanfulla*, if the repugnant discredit by ignoble lampooners continues against us and against our territory).

It is important to keep in mind that it was only upon their arrival in Brazil that these immigrants encountered an extremely racialized environment. Brazilian elites, intellectuals, and newspapers produced abundant negative stereotypes of former slaves and of poor Brazilians in general. It was this disdain for the existing population that had encouraged immigration in the nineteenth century. *Fanfulla* quickly adopted this racialized tone and negative stereotyping, whereas for these immigrants, color and race had never been an issue in Italy. It is striking to see how in July 1923, *Fanfulla* reports in great detail on a labor strike and notes of each victim whether he or she was Brazilian, Italian-Brazilian, or "black." By becoming "white," Italian immigrants found a way to distinguish themselves from the discriminated position of the Afro-Brazilians. This implied that with time, the formerly widely diverse Italian immigrant groups developed an "ethnic consciousness," and in this sense, the transformation of strangers to citizens was also a transformation of immigrants "without color," or without a colored identity, into white Brazilians.

INVENTING A TRADITION

The black population is, of course, what has been excluded from this Paulista picture. Italian immigrants came here to replace blacks and were considered superior for being white and Catholic, even if they were also manual laborers. Thus, as Lúcia Lippi argues, by being a menial worker the immigrant develops an "ethnic consciousness" neither as upper class nor as lower class.[52] In this sense, immigrants discover they are something else, ethnic, "Italian."

Rather than populating the interior of the country and working on its coffee plantations, as was originally intended, more and more immigrants remained in São Paulo. This new metropolis, full of poor immigrants and labor unions, together with modernization and the rise of a nouveau riche and middle class, created within the elite a climate of fear about foreignness and materialism, as well as nostalgia for the countryside and the lost values of the preimmigrant period. Former slaves, poor white immigrants—all these new figures needed to be accounted for in Brazilian social imaginary of that time. The rapid accumulation of capital allowed the coffee planters to adapt to the end of the slave trade by introducing modern technology to increase productivity and by gradually transitioning from slave labor to free immigrant labor. Southern European immigrants whose passages were subsidized by the São Paulo and Brazilian governments went to the economically dynamic regions of central and southern Brazil.

The city of São Paulo was quickly transformed into Brazil's leading commercial and industrial metropolis. More than 2 million European immigrants entered the state, often after receiving a free passage in exchange for work on the coffee plantations, but many also migrated to the city. Immigrants provided cheap labor for the growing factories, which, in turn, supplied expanding local markets with manufactured goods.[53] As wealthy Paulistas started doubting the cultural superiority of many of these immigrants, they became more self-assured and started to criticize the presumed superiority of Europe.

Susan Besse describes how rising industrial production from the late nineteenth to the early twentieth century increasingly replaced artisanal and household production, providing consumer goods for a growing market.[54]At the same time, women were also supposed to be anchors of stability against the destabilizing effects of industrial capitalist development, shielding the family from "corrupt" influences. One of these contradictions is Oswald de Andrade's description of modernist painter Tarsila do Amaral, whose work reflects women's ambiguous position within modernity. Andrade refers to her as "a *caipirinha* (simple Brazilian backwoods woman) dressed by—the Parisian designer—Poiret."[55]

In line with their search for a mystical "homeland" with one national race, the students of the Faculty of Law together with the anti-immigrant Nationalist League and the elite newspaper *O estado de São Paulo* erected a monument to nineteenth-century poet Olavo Bilac as a symbol of old-school nationalist culture. Furthermore, they erected another statue called *Bandeirantes*, which can still be seen in São Paulo today. Sevcenko ironically observes, "O bandeirante era apresentado como o lídimo representante das mais puras raízes sociais brasileiras, conquistador de todo o vasto sertão interior do país, pai fundador da raça e da civilização Brasileiras, em franca oposição aos 'emboabas,' pessoas estranhas à terra, traficantes desenraizados e elementos provenientes de terras estrangeiras" [56] (The bandeirante was presented as the most representative of the purest Brazilian social races, the conqueror of the whole vast backlands of the country, the founding father of the Brazilian race and civilization, in direct opposition to the new European settlers, people foreign to the land, uprooted traffickers, and elements coming from foreign lands).

DISCONTENTS: SOCIALISM AND UNIONIZATION IN THE 1920S

Poverty, migration, immigration, and unemployment helped usher in a period of radicalized politics in which protests and strikes became common in the second decade of the twentieth century. The São Paulo government actively tried to "sanitize" the city, which mainly consisted of leveling low-income residences without any consideration for where these people might go. These neighborhoods were specifically targeted, not merely for their unhygienic living conditions, but also to destroy any attempt at labor organizing. The poorest neighborhoods were those in industrial areas, inhabited by recent immigrants and former slaves. The governing elites expressed their xenophobia less through racial discrimination—after all, many of the newcomers were "whiter" than the elites themselves—than through the fear of their presence leading to the dissolution of Brazilian nationality, an internal disease that would destroy the city. It is interesting to note that the new bourgeois neighborhood Higienópolis attracted as much attention as its symbolic opposite, the immigrant, working-class neighborhood Bráz. The newspaper *O estado de São Paulo* published a whole series of articles under the title "Um bairro desprezado" ("A Neglected Neighborhood") illustrated with photographs to accentuate the poverty in the Bráz district:[57] "Bairro pobre, cuja população é na sua maioria constituída de gente simples, que mora em casas modestas, quando não habita cortiços insalubres e que se estiola nas fábricas e oficinas, o Bráz foi sempre desprezado"[58] (A poor neighborhood, whose population is mostly constituted of simple people, who live in modest houses, when not taking shelter in unhealthy *cortiços*, and which grows pale in factories in factories and offices, Bráz was always neglected). The journalist accuses the owners of the factories as well as the municipal officials of completely abandoning the neighborhood. It seems that Bráz only becomes visible on the days of *carnaval*, when wealthier families go into the neighborhood to celebrate. In an essay titled "Carnaval arquitetônico," Alcântara Machado makes the link between carnival, modernity, and race. He calls the city "uma batida arquitetônica. Tem todos os estilos possíveis e impossíveis. E todos eles brigando com o ambiente"[59] (an architectural beat. It has all possible and impossible styles. And all of them fighting with the environment). He understands that the government wants to renew and improve the city, but unfortunately for these bureaucrats "esse bonito é sempre importado. Daí o desastre estético-urbano. Lembram-se de construir uma catedral. Está certo. Mas a quem encomendam o projeto? A um alemão. E o alemão já sabe surge com uma coisa em estilo gótico. E essa coisa é aceita e está sendo feita"[60] (beauty is always imported. This is why we have an aesthetic-urban disaster. They remind themselves of building a cathedral. It's true. But who led the project? A German. And the German, everyone knows, comes up with something gothic. And this is accepted and it is being constructed). Alcântara Machado is keenly aware that this preference for foreign styles is directly related to Brazil's racial inferiority complex: "O mal é muito mais extenso do que se pode imaginar e tem origem na obsessão racial do que é estrangeiro"[61] (This evil is much more extensive than we can imagine and it originates in our racial obsession with what is foreign).

In addition to the strikes, World War I had a dramatic impact on the Brazilian economy. Since the importation of manufactured goods was impossible at this time, local industries developed out of necessity. The workers in these industries developed a new, urban "class consciousness," especially the immigrants from Italy. There were various attempts to organize these workers. In 1917, the Brazilian Socialist Party was founded, only to be dismantled in 1922 by political opponents. The Communist Party was also founded around this same time and played a major role in the São Paulo strike. In 1918 the union of textile workers amounted to around twenty thousand workers, and in that same year the União Geral dos Trabalhadores, the first national labor union, was founded.[62]

Traditionally, the educated and mainly white elite feared violence from blacks and mulattos, whom they portrayed as lazy, undisciplined, sickly, drunk, and in a constant state of vagabondage. To these fears new ones were added about disorder and violence by foreign-born factory workers, many of whom were expelled from the country on charges of being anarchists bent on overthrowing the social order. The threat of urban unrest called into question the adequacy of laissez-faire liberalism for solving social problems and suggested new roles for the state in regulating relations between workers and owners and even in intervening directly in social life. Actual progress was pitifully slow in coming, and much of the social legislation eventually passed in the 1920s was more symbolic than real, an occasion for rhetoric rather than for any serious redistribution of economic and social resources.[63] Starting with the Russian Revolution in 1917, the Communist Party begins to control labor movements, replacing previous anarchist groups.

Alcântara Machado and His Ítalo-Bandeirantes

Antônio Alcântara Machado (1901–35) was an important voice in Brazilian *Modernismo*, a movement that generally is thought to have started with The Week of Modern Art, the first in a series of events in 1922, which were symptomatic of the transformations that were taking place in Brazilian society at that time. This artistic happening can be seen as both a precursor and catalyst for the analysis of the composition of the nation that took place in the centennial year of Brazilian independence. Dawn Ades notices the break with the past and claims that the celebration of modernity was often accompanied by the reassessment of tradition. Nationalism versus

internationalism and the regional versus the centralized and cosmopolitan were key issues in Brazilian *Modernismo*. In what follows, I will analyze *Brás, Bexiga e Barra Funda*, a collection of short
stories that deals with the lives of lower-middle-class Italians in São
Paulo. I argue that the images of the Italian immigrant in these
stories are not simply stylistic devices but also references to deeply
debated questions of identity formation of that period.

Alcântara Machado published the work in 1927. The other
collection of short stories published during his lifetime, *Laranja
da China*, was published the following year, and *Vários contos*,
renamed *Contos avulsos* in later editions, was posthumously published in 1936. Alcântara Machado also worked as a theater critic
for *Jornal do comércio* and became a passionate promoter of the
modernist movement, collaborating with prominent intellectuals
such as Oswald de Andrade and Mário de Andrade, in journals
such as the *Revista de antropofagia*, which he cofounded in 1928.

Machado's work is probably best known for its use of colloquial
language, particularly that of Italian immigrants. The author uses
popular expressions, Italian phrases, and common interjections—
linguistic elements integral to contemporary speech—to describe
the lives of lower- and middle-class Italians in São Paulo. According to Sergio Milliet, the author devoted much time to riding the
streetcars of the poorer neighborhoods and frequenting cafés and
other business establishments in order to absorb the language of
the poor.[64] The author's familiarity with popular Italo-Brazilian
discourse enabled him to use this register of speech in his fiction,
thus criticizing and "modernizing" the language of his literary
predecessors.

For his part, Michael North has argued that to see linguistic
mimicry and racial masquerade simply as instances of modern
primitivism, a return to nature or a recoiling from modernity, is
to miss a far more intriguing function.[65] The real attraction of
the "other" voice is its insurrectionary opposition to the known
and familiar in language, a language opposed to the standard
one. Machado's familiarity with these popular discourses enabled
the upper-class writer to represent this register of speech in his
fiction. In line with international avant-garde movements, this

was clearly an aesthetic project. In criticizing the language of his literary predecessors, he states,

> O delírio da frase de efeito, do período clarinado, do final majes-
> toso de sinfonia italiana é o ar de família que distingue os literatos
> da terra . . . [A] arte de escrever no Brasil tem se alimentado de
> eloqüencia. Mais dia menos dia estoura de indigestão. Só se sal-
> vará com uma dieta rigorosa de caldo de simplicidade e mingau de
> discreção.[66]

> [The delirium of the proverb, the bugled period, the majestic end-
> ing of an Italian symphony is the family air that distinguishes the
> literary men from the earth . . . (The) art of writing in Brazil has
> been nourished by eloquence. But sometimes it bursts with indiges-
> tion. It will only be saved with a rigorous diet of simplicity soup and
> discretion porridge.]

Perhaps this juxtaposition of languages and the communities they signify epitomizes the belief that the division between ethnic identities is primarily linguistic and can therefore be easily over-come by the second generation, who will mostly have adopted Portuguese as their primary language. In his excellent study on Italian immigration in São Paulo, Mário Carelli notes that this depiction of Italian immigrants and their speech was important, not only for its literary innovation, but also for the role it played in overcoming the xenophobic attitude toward this community that was previously common in Brazilian fiction.[67]

The author is explicit in his cosmopolitan, anti-xenophobic attitude: in the dedication, the "Artigo de fundo," the narrator directs himself to renowned contemporary Italo-Brazilians listed on the volume's directory page. The narratives of *Bráz, Bexiga e Barra Funda* portray the varying degrees and methods of inte-gration into Paulista society achieved by Italian immigrants and especially the impact on the children on this society in trans-formation. All the twelve short stories are set in the city of São Paulo; the physical area of the city is continually brought to mind through references to specific neighborhoods, streets, and street-car lines. As indicated by the work's title, the working-class neigh-borhoods are of primary importance in these stories, providing

a veritable map of the center of the city and the three neighborhoods in the title. The stories also reflect the occupational
diversity of a city with a multifaceted economic base. In the cast
of characters, we see domestic servants, factory workers, seam-
stresses, streetcar drivers, and municipal employees.

The collection of tales is structured within the narrative framework of a newspaper that chronicles the daily life of the city's Italo-
Paulista population. The author's choice of this medium to celebrate
the role of this immigrant community in São Paulo's changing society reflects his awareness of the newspaper's importance in urban
cultures, especially in the formation of public opinion. Also, the
dynamic milieu that he presents is perfectly complemented by the
literary adaptation of the brevity, speed, and economy of journalistic discourse. The narrator, in the persona of "a redação" (the editor), declares of this work, "É um journal. Mais nada . . . Não
aprofunda . . . Em suas colunas não se encontra uma única linha
de doutrina . . . Acontecimentos de crônica urbana. Episódios de
rua"[68] (It is a newspaper. Nothing more . . . It does not examine
things carefully . . . In its columns you will not find one single line
of doctrine . . . The happenings of the urban chronicle. Episodes
from the street). The importance of newspapers is highlighted
within the stories by the frequent mention of newspapers, of *O
estado de São Paulo*, an elite publication, and of *Fanfulla*. Interestingly enough, however, Machado's work focuses on assimilation,
and many of his stories take children as their main character, which
enables a light, playful, and colloquial tone. In the story "Carmela," for example, the protagonist, a seamstress of Italian origin,
meets a boy she likes, mainly because he owns a car: "Um rapaz que
usa óculos para o Buick de propósito na esquina . . . Carmela está
interessada nele"[69] (A boy with glasses stops the Buick deliberately
at the corner . . . Carmela is interested in him). She accepts a ride
from this boy, clearly understanding that this flirt is for fun, and
her real boyfriend is for marriage. The sketch is light, brief, and
funny. Carmela's desire to improve her social status, at least tem-
porarily, is treated in Alcântara Machado's typical, slightly ironic
fashion. Nevertheless, the author never portrays the harsh lives of
factory workers or the exploitation of child labor. What would have
happened to Carmela had she been raped or had her boyfriend

found out about her innocent little flirt? Alcântara Machado is not interested in exploring these possibilities. Sometimes his continual insistence on linguistic virtuosity above any description of emotion can be almost offensive as in a scene in which a little girl is killed by an automobile. The mother is naturally extremely upset, but the others in the Palestra Itália Salon "conversam imaginando como o *Fanfulla* daria a notícia da morte da menina. Percebe-se que foi um sujeito rico quem a atropelou. As pessoas não acreditam sequer que o jornal dê a notícia como devea ser dada, pelo fato do rapaz ser rico, poderoso . . . Filho de rico manda nesta terra. Pode matar sem medo. É ou não é, seu Zamponi?"[70] ([They] converse imagining how *Fanfulla* would redact the story of the young girl's death. They noticed that it was a rich man who ran over her. The people don't even believe that the paper would print the story as it should be written, because of the fact that the man was rich, powerful: "The rich man's son is in charge in this land. He can kill without fear. Isn't that right, Mr. Zamponi?").

In this fictional universe, social injustice and labor struggles are totally absent, even irrelevant. The frequently denigrating and patronizing attitude of the writer seemingly contradicts his explicit purpose of celebrating Italian immigrant culture. It seems to me that this is an unavoidable outcome of the *Modernista* aesthetic. The topic of racial miscegenation should be seen as an integral part of Brazilian *Modernismo* and in "Anthropophagy." Anthropophagy is one of the most original theses among those formulated in Latin America with the purpose of solving the tensions and contradictions typical of a country that, on one hand, wanted to cut loose from its colonized roots, and, on the other, was trying to keep up with the revolutionary artistic and cultural manifestations of the European historical vanguards. Heloísa Buarque de Hollanda has made a valuable comparison between official racial policies and the paternalist attitude of many *Modernista* writers, such as Alcântara Machado.[71] She argues that the model of *Antropofagia*, instead of establishing a direct confrontation between different cultures, proposes to "eat the differences up"—that is, to assimilate by force. This eating of the other is followed by the option of either expelling or ignoring:

O Brasil, "deglutidor das diferenças," construído pelo modernismo como o reino da "cordialidade," do calor receptivo, da predis-
posição "nata" para receber o "outro" e com ele se identificar. Neste
mesmo caminho, reforço agora a idéia da dificuldade em estabelecer
e instrumentalizar com clareza, . . . as divergências de interesses
entre classes sociais ou entre grupos étnicos ou sexuais no Brasil.[72]

[Brazil, "glutton of differences," built by *modernismo* as the kingdom of cordiality, of warmwelcomes, of the "natural" predisposition to receive an identity with the "other." In this same vein, I
reiterate the idea of the difficulty in establishing and instrumental-
izing with clarity . . . the diverging interests between social classes
or between ethnic groups or sexes in Brazil.]

In his introduction, Alcântara Machado develops his own theory
of racial miscegenation, influenced by the ideological debates and
local politics of his time. He summarizes the history of the ethnic
formation of Brazilian society under what Chalmers calls "o prisma
da teoria da miscigenação emprestado ao *Manifesto Pau-Brasil* de
Oswald de Andrade"[73] (the prism of the theory of miscegenation
lent to the Manifesto Pau-Brasil by Osvald de Andrade). After
proclaiming this work a piece of journalism, the writer immediately mentions the city's increasingly complex racial makeup and
comments, "O aspecto étnico-social dessa novíssima raça de gigan-
tes encontrará amanhã o seu historiador. E será então analisado
e pesado num livro"[74] (The ethnic-social aspect of this extremely
new race of giants will tomorrow find its historian. And it will
then be analyzed and weighed in a book). Interestingly enough,
he calls these new citizens "novos mamalucos" (new half-breeds)
and "estrangeiros paulistanizados" (Paulistified strangers). The
first group of mamelucos were those born of the Portuguese and
the indigenous population, and the second were those born of the
slaves and their masters. The third group of mamelucos, according
to the author, are the immigrants: "Do consórcio da gente imigrante
com o ambiente, do consórcio da gente imigrante com a indígena
nasceram os novos mamalucos. Nasceram os *italianinhos* (sic [in
original]). O Gaetaninho. A Carmela. Brasileiros e paulistas. Até
bandeirantes"[75] (Of the type of immigrants with that environment,
of the type of immigrants with the indigenous population, were

born the new mamelucos. The little Italians were born. The little Gaetano. Carmela. Brazilians and Paulistas. Even bandeirantes).

The importance of the Italian immigrants in the formation and direction of contemporary São Paulo is underscored when the author compares them to the famed *bandeirantes* of the seventeenth century, Paulista adventurers who explored the hinterlands in search of land, riches, and indigenous slave labor. The Italo-Brazilians are urban *bandeirantes*, penetrating and thereby restructuring traditional Paulista society, which is representative of the new frontier of a nation becoming more urban than rural.

WHITENESS OF A DARKER COLOR

I TAKE THE TITLE OF THE CONCLUSION OF MY book from *Whiteness of a Different Color: European Immigrants and the Alchemy of Race* by Matthew Frye Jacobson. Whiteness in South America had its own developments and was as intensely debated as in the United States. In general, comparison with the United States limits the consideration of race in Latin America or, in this case, Argentina and Brazil. For this reason, I have approached the idea of racial democracies from an endogenous perspective—by incorporating all groups in the Brazilian or Argentine population relative to each other, rather than using the United States as the measure of comparison—in hopes of questioning and contributing to existing scholarship.

In the realm of culture, I also think tales of transatlantic crossings have hardly ceased to be relevant, as can be seen for instance in the excellent Argentine film *Bolivia* (Adrián Caetano, 2002), which ferociously denounces *Porteño* racism and the exploitation of illegal immigrant workers by Argentines of European origins, or Paula Hernández's nostalgic *La Herencia* (2002), which I saw with cheering Italo-Paulistas in São Paulo and which details the journal of an Italian immigrant back to her country of birth. Alicia Steimberg and Mempo Giardinelli are well-established writers who have traced their family histories in their fiction, and with *El infierno prometido* (2006) Elsa Drucaroff has written a beautiful novel about Jewish immigration and prostitution in Buenos Aires. The Brazilian interest in ethnicity has always caught my attention during my travels to this country, from people who proudly mentioned their own

Dutch origins to me, to the recent renovation of São Paulo's *Museu de Inmigrante* (Museum of Immigrants). Numerous films, magazines, and documentaries address the era of mass immigration, and one has only to think about well-respected writers such as Milton Hatoum, Moacyr Scliar, and Raduan Nasser in Brazil to realize that, by recognizing the importance of ethnicity within Latin America, there are many more research paths to be explored.

In the Southern Hemisphere of the American continent, with the threat of being viewed as "black" always looming, European immigration was seen as a crucial agent of modernization by both Argentinean and Brazilian elites and linked to notions of cultural as well as economic and political progress. Some intellectuals and politicians sought "pure" European immigrants who would recreate the Old World in the New. Others were more pessimistic and argued that the reality of miscegenation underneath the fiction of racial democracy prevented the South American countries from achieving their rightful place among elite nations. Rapid urbanization, industrialization, immigration, and demographic growth had restructured society, creating new problems that the traditional oligarchies had failed to address. Immigrants, then, were imbued with many layers of meaning, a composite whose boundaries were unstable and constantly shifting, even as they discover significant elements of continuity across the differentials of period and context.

In this study I have shown that, while the process of mass immigration was similar historically, the reception of immigrants in Argentina and Brazil was quite different. Both countries changed dramatically, however, and started inventing a "national character" precisely as a result of mass immigration and the subsequent doubt over whether the nation could eventually be homogenized. In Argentina, where racial differences were not always clearly visible, immigration issues were translated into or affiliated with class. In Brazil, on the other hand, immigrants were categorized using various racial labels while a cosmopolitan *mestiçagem* was eventually installed as an official discourse replacing earlier deterministic ideas on race, thus providing the institution that would provide a Brazilian "national type."

One of the merits of Jeffrey Lesser's *Negotiating National Identity* is that it criticizes the still too often undisputed "triangle theory" in which Brazil is seen as a society created from the "collision of three races," Africans (blacks), Europeans (whites), and Indians (indigenous). In this theory, the mixture of peoples found within the area enclosed by the border of the triangle created infinite genetic possibilities. Thus, according to this traditional paradigm, Brazil is a country struggling with an identity that always exists at some point along the continuum. Many academics have presumed or implied that anyone without African or indigenous ancestry belongs, by definition, in the category of "white." The flaw of the triangle theory is, of course, the danger of not interrogating these fixed, anachronistic categories of "black" and "white," while failing to notice that what it actually meant to be "white" shifted markedly over time.

Probing the intersection of immigrant ethnicity and national identity allows a more detailed picture to emerge of how immigrants, on the one hand, and elites, on the other, reacted when confronted with a changing world that demanded new responses to cultural questions. As a result of mass immigration, both groups came to agree that the American countries were engaged in a mythic search for a national race and character. Although the period of mass immigration from Europe and eugenic beliefs in racial hierarchies is long over, it seems that immigration issues continue to be of interest in contemporary Latin America. In this respect, the large migrations from the Brazilian Northeast to São Paulo that are rapidly transforming Brazilian society are telling. In addition, due to the persistent economic crises in Latin America and globally, many Argentines as well as Brazilians are rethinking their relationship with Europe. Ethnicity suddenly became crucial when people were able to prove their European origin in order to apply for a passport and work in the European Union. Ultimately, where we come from, who we are allowed to be, and why, are all questions that seem to be asked by more people than ever in this era of globalization.

NOTES

INTRODUCTION

1. Tulio Halperín Donghi, "¿Para qué la inmigración? Ideología y política inmigratoria en la Argentina (1810–1914)," in *El espejo de la historia. Problemas argentinos y perspectivas latinoamericanas* (Buenos Aires: Sudamericana, 1987), 189–238.
2. Quoted in Halperín Donghi, "¿Para qué la inmigración?" 196. Unless otherwise noted, all translations are my own. For more information about this phase of the invention of nation and state, see Nicholas Shumway, *The Invention of Argentina* (Berkeley: University of California Press, 1991), 1–23, 81–111; Tulio Halperín Donghi, "Argentine Counterpoint: Rise of the Nation, Rise of the State," in *Beyond Imagined Communities: Reading and Writing the Nation in Nineteenth-Century Latin America*, ed. Sara Castro-Klarén and John Charles Chasteen (Baltimore, MD: Johns Hopkins University Press, 2003), 33–53.
3. Matthew Frye Jacobson, *Whiteness of a Different Color* (Cambridge, MA: Harvard University Press, 1998), 6.
4. Ibid., 7.
5. Caren Kaplan, *Questions of Travel: Postmodern Discourses of Displacement* (Durham, NC: Duke University Press, 2000), 31.
6. Roberto Schwarz, "Nacional por subtração," in *Que horas são?* (São Paulo: Companhia das Letras, 1987), 29–48.
7. Jeffrey Lesser, *Negotiating National Identity: Immigrants, Minorities, and the Struggle for Ethnicity in Brazil* (Durham, NC: Duke University Press, 1999), 1–12.
8. José Moya, *Cousins and Strangers* (Berkeley: University of California Press, 1998).
9. Nancy Stepan, *The Hour of Eugenics: Race, Gender, and Nation in Latin America* (Ithaca, NY: Cornell University Press, 1991), xiv–xix.
10. Ibid., 136–37.
11. See Michel Foucault, *"Society Must Be Defended": Lectures at the Collège de France, 1975–1976* (London: Penguin, 2003), esp. Lecture 3, January 21, 1976; and *The History of Sexuality: An Introduction* (New

York: Vintage, 1988), esp. 17–49. See also Ann Stoler, *Race and the Education of Desire Race and the Foucault's History of Sexuality and the Colonial Order of Things* (Durham, NC: Duke University Press, 1995), for a detailed discussion on Foucault and race.

12. Holloway, 7; As Holloway explains, for obvious reasons, historians of Brazil have often focused on the institution of slavery in their work.

13. Graham Burchell, Collon Gordon, and Peter Miller, eds., *The Foucault Effect: Studies in Governmentality* (Chicago: University of Chicago Press, 1991), 5.

14. Georg Simmel,"The Stranger," in *On Individuality and Social Forms* (Chicago: University of Chicago Press, 1971), 143–49.

13. Bonnie Honig, "Ruth, the Model Émigré: Mourning and the Symbolic Politics of Immigration," in *Cosmopolitics: Thinking and Feeling Beyond the Nation*, ed. Cheah Pheng and Bruce Robbins (Minneapolis: Minneapolis University Press, 1998), 192–215.

16. Edward Said, *The World, the Text, and the Critic* (Cambridge, MA: Harvard University Press, 1983), 1–30.

17. Caren Kaplan, *Questions of Travel*, 110.

18. James Clifford, "Diasporas," *Cultural Anthropology* 9, no. 13 (1994): 321.

19. Lisa Lowe, introduction to *Immigrant Acts: On Asian American Cultural Politics* (Durham, NC: Duke University Press, 1996), 3.

20. Julyan Peard, *Race, Place, and Medicine: The Idea of the Tropics in Nineteenth-Century Brazilian Medicine* (Durham, NC: Duke University Press, 1999), 84.

21. João Cardoso de Menezes e Souza, *Theses sobre colonização do Brasil: projecto de solução as questões sociais, que se prendem à este dificill problema* (Rio de Janeiro: Typografia nacional, 1875).

22. Raymundo Nina Rodrigues, *As raças humanas e a responsabilidade penal no Brasil* (São Paulo: Companhia Editora Nacional, 1938).

23. Francine Masiello, *Lenguaje e ideología. Las escuelas argentinas de vanguardia* (Buenos Aires: Hachette, 1986), 37.

24. Arrigo De Zettiry, *Manual del emigrante italiano* (Buenos Aires: Centro Editor de América Latina, 1983).

CHAPTER 1

1. Jeffrey Lesser, "The Hidden Hyphen," in *Negotiating National Identity: Immigrants, Minorities, and the Struggle for Ethnicity in Brazil* (Durham, NC: Duke University Press, 1999), 6.
2. Ibid., 7.
3. Nancy Stepan, *The Hour of Eugenics: Race, Gender, and Nation in Latin America* (Ithaca, NY: Cornell University Press, 1991), 44.
4. Joel Outtes, "Disciplining Society through the City: The Genesis of City Planning in Brazil and Argentina (1894–1945)," *Bulletin of Latin American Research* 22, no. 2 (2003): 151.
5. Stepan, *The Hour of Eugenics*, 45.
6. Quoted in ibid., 45.
7. Dain Borges, "'Puffy, Ugly, Slothful and Inert': Degeneration in Brazilian Social Thought, 1880–1940," *Journal of Latin American Studies* 25 (1993): 241.
8. Julyan Peard, *Race, Place, and Medicine: The Idea of the Tropics in Nineteenth-Century Brazilian Medicine* (Durham, NC: Duke University Press, 1999), 92–95.
9. João Cardoso de Menezes e Souza, *Theses sobre colonização do Brasil: Projeto de solução as questões sociais, que se prendem à este difícil problema* (Rio de Janeiro: Typografia nacional, 1875), 4.
10. Menezes e Souza, quoted by Lesser, *Negotiating National Identity*, 7.
11. Menezes e Souza, *Theses*, 4.
12. Ibid., 4.
13. Ibid., 5.
14. Thomas Holloway, *Policing Rio de Janeiro: Representation and Resistance in a 19th-Century City* (Stanford, CA: Stanford University Press, 1993), 4–5.
15. Ibid., 7.
16. Raymundo Nina Rodrigues, *As raças humanas e a responsabilidade penal no Brasil* (São Paulo: Companhia Editor Nacional, 1938). See also the discussion by Ricardo Salvatore, "Penitentiaries, Visions of Class, and Export Economies: Brazil and Argentina Compared," in *The Birth of the Penitentiary in Latin America : Essays on Criminology, Prison Reform, and Social Control, 1830–1940*, ed. Ricardo Salvatore and Carlos Aguirre (Austin, TX: University of Texas Press: 1996), 194–223.
17. For more on his family background and strangely ambiguous interest in Afro-Brazilian culture, see Thomas Skidmore, *Black into White: Race*

and Nationality in Brazilian Thought (Durham, NC: Duke University Press, 1995), 57–62.

18. Peard, *Race, Place and Medicine*, 102–4.

19. Eduardo Prado, *A ilusão Americana* (São Paulo: Editora Brasiliense, 1961), 84. The italics and English "by any means" are in the original.

20. Ibid., 113.

21. Ibid., 115.

22. Ibid., 122–23.

23. Ibid., 79.

24. Ibid., 19.

25. Ibid., 14.

26. Ibid., 131.

27. Ibid.

28. Ibid., 175.

29. Ibid., 176–77.

30. Ibid., 133.

31. Ibid., 134.

32. Ibid., 129.

33. Ibid., 171.

34. Holloway, *Policing Rio de Janeiro*, 274.

35. Ibid., 287.

36. Outtes, "Disciplining Society," 140.

37. Holloway, *Policing Rio de Janeiro*, 23–24.

38. Ibid., 24.

39. Gilberto Freyre, *Sobrados e mucambos; decadência do patriarcado rural e desenvolvimento do urbano*, 4 ed. (Rio de Janeiro: José Olympio, 1968), 607.

40. Artur Azevedo, "Flocos," *Correio do Povo* (Rio de Janeiro), May 18, 1890. Also quoted in Jean-Yves Mérian, *Aluísio Azevedo: Vida e obra. O verdadeiro Brasil do século XIX* (Rio de Janeiro: Espaço e tempo, 1988), 517.

41. Pardal Mallet, "O cortiço," *Gazeta de Notícias* (Rio de Janeiro), May 25, 1890. See also Mérian, *Aluísio Azevedo*, 518.

42. Amy Chazkel, "The Crônica, the City, and the Invention of the Underworld: Rio de Janeiro, 1889–1922," *Estudios interdisciplinarios de América Latina y el Caribe* 12, no. 1 (2001): 79–105.

43. Quoted in Mérian, *Aluízio Azevedo*, 97–98.

44. Outtes, "Disciplining Society," 159.

45. *O Corsário* (Rio de Janeiro), April 26, 1883, quoted in Jean-Yves Mérian, *Aluízio Azevedo*, 516.

46. Gladys Sabina Ribeiro, *Mata galegos: os portugueses e os conflitos de trabalho na República Velha* (São Paulo: Brasiliense, 1989), 15.

47. Quoted in Luiz Felipe de Alencastro and Maria Luiza Renaux, "Caras e modos dos migrantes e imigrantes," in *História da vida privada no Brasil*, vol. 2 of *Império: a corte e a modernidade nacional*, ed. Fernando Novais (São Paulo: Compania de Letras, 1997), 310. The term *"galego"* in the original refers technically to a person from Galicia, Spain, but is used figuratively here to refer pejoratively to a foreigner of low socioeconomic status. Similarly, *cangueiro* is a term literally meaning one who bears a *canga* or yoke but is here used figuratively to refer to a lazy or negligent individual.

48. Ribeiro, *Mata galegos*, 31.

49. Ibid., 28.

50. Ibid., 15.

51. Ibid., 27.

52. Aluísio de Azevedo, *O cortiço*, http://www.cervantesvirtual.com/servlet/SirveObras/05815063290570695209079/index.htm. Chapter 8. Translated by David H. Rosenthal as *The Slum* (London: Oxford University Press, 2000), 70. In this and subsequent citations of *O cortiço*, I indicate first the chapter of the Portuguese online edition and then the published English translation.

53. Azevedo, *O cortiço*, VII; *The Slum*, 55.

54. Azevedo, *O cortiço*, XVIII; *The Slum*, 168.

55. Azevedo, *O cortiço*, I; *The Slum*, 5.

56. Azevedo, *O cortiço*, I; *The Slum*, 2.

57. Azevedo, *O cortiço*, I; *The Slum*, 9.

58. Azevedo, *O cortiço*, I; *The Slum*, 4.

59. Azevedo, *O cortiço*, I; *The Slum*, 35.

60. Ibid.

61. Azevedo, *O cortiço*, VII; *The Slum*, 61.

62. Azevedo, *O cortiço*, VII; *The Slum*, 178.

63. Azevedo, *O cortiço*, IX; *The Slum*, 75.

64. Elizabeth Marchant, "Naturalism, Race, and Nationalism in Aluísio Azevedo's *O mulato*," *Hispania* 83, no. 3 (2000): 445–53.

65. Rita Felski, *The Gender of Modernity* (Cambridge, MA: Harvard University Press, 1995), 40–41.

66. Anne McClintock, "Family Feuds: Gender, Nationalism, and the Family," *Feminist Review* 44 (1993): 61–80.

67. For a more detailed discussion on this matter, see Felski, *The Gender of Modernity*, 39–40.
68. About the change in Júlia Lopes de Almeida's literary reputation, see Peggy Sharpe, "O caminho crítico d'A viúva Simões," in *A viúva Simões* (Florianópolis: Editorial Mulheres, 1999), 9–26.
69. João do Rio, "Um lar de Artistas," in *Momento Literário* (Rio de Janeiro: Fundação Biblioteca Nacional, 1994), http://www.cervantesvirtual .com/servlet/SirveObras/12693858724598273098435/p0000001 .htm#I_5.
70. Mary Louise Pratt, "Women, Literature, and the National Brother-hood," *Nineteenth-Century Contexts* 18, no. 1 (1994): 27–47.
71. João do Rio, "Um lar de Artistas."
72. Ibid.
73. Ibid.
74. Oral communication by Nadilza Moreira made at the Brazilian Studies Association conference at Tulane University, March 29, 2008. Moreira is currently examining the family archive in Rio de Janiero owned by Almeida's grandson Carlos de Almeida and is preparing a biography of Almeida.
75. Peggy Sharpe, "O caminho crítico d'A viúva Simões," in *A viúva Simões* (Florianópolis: Editorial Mulheres, 1999), 20.
76. Jeffrey Needell, *A Tropical Belle Époque: Elite Culture and Society in Turn-of-the-Century Rio de Janeiro* (Cambridge: Cambridge University Press, 1987), 126.
77. Almeida, *A casa verde*, 139.
78. Sônia Roncador correctly observes, "[O] lar . . . é um laboratório de formação do caráter nobre dos futuros homens e mulheres de bem da nação" (the home . . . is a formative laboratory of the noble character of the future respectable men and women of the nation). "O demônio familiar: lavadeiras, amas-de-leite e criadas na narrativa de Júlia Lopes de Almeida," *Luso-Brazilian Review* 44, no. 1 (Spring 2007): 94.
79. Louise Guenther, "The British Community of 19th Century Bahia: Public and Private Lives" (working paper, University of Oxford Cen-tre for Brazilian Studies Working Paper Series, http://www.brazil.ox.ac .uk/__data/assets/pdf_file/0003/9399/Guenther32.pdf).
80. Robert Dundas, *Sketches of Brazil, Including New Views on Tropical and European Fever with Remarks on a Premature Decay of the System Inci-dent to Europeans on Their Return from Hot Climates* (London: John Churchill, 1852), quoted in Guenther, "The British Community of 19th Century Bahia," 20.
81. In *A viúva Simões*, Sara, the daughter and amorous rival of her own mother is part German; in *A família Medeiros*, the heroine, Eva, is

educated in German. The results for Almeida are the same: healthy athletic women who are not afraid of hard work and are capable of combining physical beauty with intelligence.

82. Maria Graham, *Journal of a Voyage to Brazil, and Residence There during Part of the Years 1821, 1822, 1823* (London: Longman, Hurst, Rees, Orme, Brown, and Green, 1824). Maria Graham was the thirty-six-year-old wife of a British naval officer, who was assigned to patrol the ports of Brazil during its independence battles. See Graham, *Journal*, 135–42. Quoted in Guenther, "The British Community of 19th Century Bahia," 21.

83. Almeida, *A casa verde*, 23.

84. Ibid., 8–9.

85. Ibid., 23.

86. Ibid., 140.

CHAPTER 2

1. Arnd Schneider, "The Transcontinental Construction of European Identities: A View from Argentina," *Anthropological Journal on European Cultures* 5, no. 1 (1996): 95–105.

2. Tulio Halperín Donghi, "Argentine Counterpoint: Rise of the Nation, Rise of the State," in *Beyond Imagined Communities: Reading and Writing the Nation in Nineteenth-Century Latin America*, ed. Sara Castro-Klarén and John Charles (Baltimore, MD: Johns Hopkins University Press, 2003), 38n2.

3. Carl Solberg, *Immigration and Nationalism, Argentina and Chile 1890–1914* (Austin: University of Texas Press, 1970), 110–16.

4. William Acree, "Tracing the Ideological Line: Philosophies of the Argentine Nation from Sarmiento to Martínez Estrada," *A contracorriente* 1, no. 1(2003):102.

5. Etiénne Balibar and Immanuel Wallerstein, *Race, Nation, Class* (London: Verso, 1991), 10.

6. Halperín Donghi, "Argentine Counterpoint," 53.

7. Beatriz Sarlo,"Modernidad y mezcla cultural. El caso de Buenos Aires," in *Modernidade: Vanguaras artísticas na América Latina*, ed. Ana Maria de Moraes Belluzzo (São Paulo: Fundação Memorial da América Latina, 1990), 31–43.

8. Beatriz Sarlo, *Una modernidad periférica* (Buenos Aires: Nueva Visión, 1988), 27.

9. Marisella Svampa, "Immigración y nacionalidad," *Studi Emigrazioni/Études Migrations* 30 (1993): 297.

10. Lilia Ana Bertoni, "La naturalización de los extranjeros 1887–1893: ¿Derechos políticos o nacionalidad?" *Desarrollo económico* 32, no. 125 (1992): 57–77. See especially pages 58–64.

11. Juan Suriano, *La huelga de inquilinos de 1907* (Buenos Aires: Center for the Economic Analysis of Law, 1983), 19.

12. Solberg, *Immigration*, 116.

13. Julio Ramos, "Faceless Tongues: Language and Citizenship in Nineteenth-Century Latin America," in *Displacements: Cultural Identities in Question*, ed. Angelika Bammer (Bloomington: Indiana University Press, 1994), 27.

14. Gladys Onega, *La inmigración en la literatura argentina: 1880–1910* (Buenos Aires: Centro Editor de América Latina, 1982), 9–10.

15. Halperín Donghi, "¿Para qué la inmigración?" 222.

16. Francine Masiello, *Between Civilization and Barbarism: Women, Nation, and Literary Culture in Modern Argentina* (Lincoln: University of Nebraska Press, 1992), 37.

17. Ibid.

18. José María Miró [Julián Martel, pseudo.], *La bolsa* (Buenos Aires: Imprima, 1979), 28.

19. Georg Simmel, "The Metropolis and Mental Life," in *On Individuality and Social Forms*, ed. Donald N. Levine and Morris Janowitz (Chicago: University of Chicago Press, 1971), 330.

20. Marshall Berman, *All That Is Solid Melts into Air* (New York: Penguin, 1988), 111.

21. Martel, *La bolsa*, 31.

22. Beatriz Sarlo, "Oralidad y lenguas extranjeras. El conflicto en la literatura argentina durante el primer tercio del siglo XX," in *Oralidad y argentinidad: estudios sobre la función del lenguaje hablado en la literatura argentina*, ed. Walter Bruno Berg and Markus Klaus Schaffauer (Tübingen, Germany: Narr, 1997), 28–41.

23. Martel, *La bolsa*, 42.

24. Ibid., 26.

25. Ibid.

26. Gabriela Nouzeilles, "Pathological Romances and National Dystopias in Argentina Naturalism," *Latin American Literary Review* 24, no. 47 (1996): 5; Doris Sommer, *Foundational Fictions: The National Romances of Latin America* (Berkeley: University of California Press, 1991), 30–51.

27. Masiello, *Lenguaje*, 14.

28. Martel, *La bolsa*, 59.

29. Ibid., 246.

30. Rita Felski, *The Gender of Modernity* (Cambridge, MA: Harvard University Press, 1995), 63. (See chapter 1, note 63.)

31. Martel, *La bolsa*, 38.

32. Hugo Vezzetti, *La locura en la Argentina* (Buenos Aires: Folios, 1983), 19.

33. Terán, *Positivismo y nación*, 19. For more information, see Gustave Le Bon, *Psychologie des foules* (Paris: Presses universitaires de France, 1991).

34. Ramos Mejía, *Las multitudes argentinas*, 15.

35. Ibid., 16.

36. Ibid., 17.

37. Ibid., 18

38. Ibid., 19. Gongorism, in the original "Gongorismo," refers to the Spanish Baroque poet Luis de Góngora y Argote (1561–1627), famous for his ornate lyrical style, also known as "culturanismo," which is referenced here by Ramos Mejía as an insult, a needless complication of a particular issue.

39. Karen Mead, "Gendering the Obstacles to Progress in Positivist Argentina 1880–1920," *Hispanic American Historical Review* 77, no. 4 (1997): 645.

40. Ramos Mejia, *Las multitudes argentinas*, 153.

41. Ibid., 153.

42. Ibid., 155–56.

43. Ibid., 157.

44. Ibid., 160.

45. Ibid., 162.

46. Ibid., 164.

47. Ibid., 169.

48. Ibid., 182.

49. Ibid., 184–85.

50. Anne McClintock, "Family Feuds: Gender, Nationalism, and the Family," *Feminist Review* 44 (1993): 63. (See Chapter 1, note 60.)

51. Bonnie Frederick, *Wily Modesty: Argentina Women Writers, 1860–1910* (Tempe: Arizona State University Press, 1998), 31.

52. Masiello, *Between Civilization and Barbarism*, 14.

53. Simmel, "The Metropolis and Mental Life," 143.

54. Emma de la Barra de Llanos [César Duayen, pseudo.], *Stella (vida de costumbres argentinas)* (Barcelona: Editorial Maucci, 1908), 82.

55. Ibid., 26.

56. Benedict Anderson, *Imagined Communities* (London: Verso, 1991), 143.

57. de la Barra de Llanos, *Stella*, 163.

58. Ibid., 149.
59. Ibid., 171.
60. Ibid., 39.
61. Ibid., 42.
62. Ibid., 51.
63. Ibid.
64. Ibid., 55.
65. Ibid., 236.
66. Sylvia Molloy, "Madre Patria y la madrastra," *La Torre* 1, no. 1 (1987): 46. See also Christina Civantos, *Between Argentines and Arabs: Argentine Orientalism, Arab Immigrants and the Writing of Identity* (Albany: State University of New York Press, 2006), especially 31–51.
67. de la Barra de Llanos, *Stella*, 192. Christiana (or Kristiana) is Norway's capital city, now known as Oslo.
68. Ibid., 179.
69. Ibid., 225.
70. Ibid., 221.
71. Ibid., 237.
72. Ibid., 239.
73. Bonnie Honig, "Ruth, the Model Émigré," in *Cosmopolitics: Thinking and Feeling Beyond the Nation*, ed. Cheah Pheng and Bruce Robbins (Minneapolis: Minneapolis University Press, 1998), 192–215.
74. de la Barra de Llanos, *Stella*, 60.

CHAPTER 3

1. Carlos Altamirano, "La fundación de la literature argentina," in *Ensayos argentines de Sarmiento a la vanguardia*, ed. Carlos Altamirano and Beatriz Sarlo (Buenos Aires: Centro Editor de América Latina, 1983), 114.
2. Scholarly treatments of this period tend to emphasize the xenophobic nature of Argentine cultural nationalism. See, for example, Carl Solberg, *Immigration and Nationalism in Argentina and Chile* (Austin: University of Texas Press, 1970), especially 149–50. Also see Richard Slatta, "The Gaucho in Argentina's Quest for Identity," *Canadian Review of Studies in Nationalism* 12 (1985): 98–122; David Viñas, "'Niños' y 'criados favoritos': De *Amalia* a Beatriz Guido a través de *La gran aldea*," in *Literatura argentina y realidad política* (Buenos Aires: Centro Editor de América Latina, 1982), 78–112, especially 93–106; and Josefina Ludmer, *El género gauchesco: un tratado sobre la patria* (Buenos Aires: Sudamericana, 1988).
3. William Acree, "Tracing the Ideological Line," *A contracorriente* 1, no. 1 (2003): 104, prefers the term "nativism." I, however, have opted to use the term "cultural nationalism," which Ricardo Rojas used.

4. Samuel Baily, "Las sociedades de ayuda mutual y el desarrollo de una comunidad italiana en Buenos Aires 1858–1918," *Desarrollo economico* 21, no. 84 (1982): 485–514.
5. Diego Armus, *Manual del emigrante italiano* (Buenos Aires: Centro Editor de América Latina, 1983).
6. Ibid., 15.
7. Beatriz Sarlo and Carlos Altamirano, "La Argentina del centenario; Campo intelectual, vida literaria y temas ideológicos" and "La fundación de la literatura argentina," in *Ensayos argentinos: De Sarmiento a la vanguardia*, ed. Carlos Altamirano and Beatriz Sarlo (Buenos Aires: Espasa Calpa, 1997), 161–200, 201–10.
8. Sarlo, "La Argentina del centenario," 72.
9. Jeane Delaney, "Making Sense of Modernity: Changing Attitudes toward the Immigrant and the Gaucho in Turn-of-the-Century Argentina," *Comparative Studies in Society and History* 38, no. 3 (1996): 439.
10. Sarlo, "La Argentina del centenario," 72.
11. Ibid., 74.
12. A clear summary of ideas from this period is included in the second chapter of José Luis Romero, *El desarrollo de las ideas en la Argentina del siglo XX* (Mexico: Fondo de Cultura Económica, 1965. Also see the previously cited articles by Sarlo and Altamirano.
13. Francis Korn, "Algunos aspectos de la asimilación de immigrants en Buenos Aires," in *Los fragmentos del poder*, ed. Torcuato di Tella (Buenos Aires: Jorge Álvarez, 1969), 452.
14. Ibid., 456–57.
15. Tulio Halperín Donghi, "¿Para qué la inmigración? Ideología y política inmigratoria y aceleración del proceso modernizador: El caso argentino." In *El espejo de la historia. Problemas argentinos y perspectivas latinoamericanas* (Buenos Aires: Sudamericanas, 1987), 214. (See introduction, note 1.)
16. Acree, "Tracing the Ideological Line," 104.
17. Scholars have disagreed over the reasons for this low naturalization rate. See, for example, Halperín Donghi, "¿Para qué la inmigración," 464–65; Torcuato di Tella, "El impacto"; and Hilda Sábato, *The Many and the Few: Political Participation in Republican Buenos Aires* (Stanford, CA: Stanford University Press, 2001).
18. Jeane Delaney, "National Identity, Nationhood, and Immigration in Argentina: 1810–1930," *Stanford Electronic Humanities Review* 5, no. 2 (1997), http://www.stanford.edu/group/SHR/5–2/delaney.html#1.
19. Ibid.

20. For a complete biography, see Earl Glaubert, "Ricardo Rojas and the Emergence of Argentine Cultural Nationalism," *Hispanic American Historical Review* 43, no. 1(1963): 1–13.

21. For a helpful discussion of the backlash against positivism and how it helped shape the emerging nativism of the period, see Rock, "Intellectual Precursors of Conservative Nationalism in Argentina, 1900–1927," 271–300, especially 272–75.

22. David Rock, "From the First World War to 1930," in *Argentina Since Independence*, ed. Leslie Bethell (Cambridge: Cambridge University Press, 1993), 139–72.

23. Solberg, *Immigration and Nationalism, Argentina and Chile 1890–1914*, 110–16.

24. Diana Sorensen, "Ricardo Rojas, lector del Facundo: Hacia la construcción de la cultura nacional," *Filología* 21, no. 1 (1986): 173–81.

25. Adriana Puiggrós, *Sujetos, disciplina y curriculum: En los orígenes del sistema educativo argentine* (Buenos Aires: Galerna, 1990), 79–81.

26. Ricardo Rojas, *La restauración nacionalista* (Buenos Aires: Ministerio de justicia e instrucción pública, 1922), 9–16.

27. Ibid., 10.

28. Ibid., 14–15.

29. Quoted in Julio Ramos, "Faceless Tongues: Language and Citizenship in Nineteenth-Century Latin America," in *Displacements: Cultural Identities in Question*, ed. Angelika Bammer (Bloomington: Indiana University Press, 1994), 25–46.

30. Jeane Delaney, "Making Sense of Modernity: Changing Attitudes toward the Immigrant and the Gaucho in Turn-of-the Century Argentina," *Comparative Studies in Society and History* 38, no. 3 (1996): 435.

31. Rojas, *La restauración nacionalista*, 120.

32. Ibid., 136–37.

33. For more detailed information on this matter, see Chapter 3, "La function política de la educación," in Juan Carlos Tedesco, *Educación y sociedad en la Argentina: 1880–1945* (Buenos Aires: Ediciones Solar, 1986), 63–88.

34. Puiggrós, *Sujetos, disciplina y curriculum*, 38–46.

35. Beatriz Sarlo, *Una modernidad periférica* (Buenos Aires: Nueva Visión, 1988), 245.

36. Beatriz Sarlo, "Oralidad y lenguas extranjeras," in *Oralidad y argentinidad: estudios sobre la función del lenguaje hablado en la literatura argentina*, ed. Walter Bruno Berg and Markuys Klaus Schaffauer (Tübingen, Germany: Narr, 1997), 27.

37. Víctor A. Mirelman, *Jewish Buenos Aires, 1890–1930: In Search of an Identity* (Detroit, MI: Wayne State University Press, 1990), 15.

38. Senkman gives the following statistics: In 1896, there were five colonies in the provinces of Buenos Aires, Entre Ríos, and Santa Fe, with 6.757 colonists and their families. The total land bought by de Hirsch's Jewish Colonization Association was at its peek more than 600 hectares. Senkman does not mention how long the colonizers managed to survive nor how many, like Gerchunoff's family, decided to return to the city.

39. Naomi Lindstrom, *Jewish Issues in Argentine Literature. From Gerchunoff to Szichman* (Columbia: University of Missouri Press, 1989), 6.

40. Ibid.

41. I have used the English translation: Prudencio de Pereda, trans., *The Jewish Gauchos of the Pampas* (Alburquerque: University of New Mexico Press, 1998).

42. Ibid., xxx.

43. Ibid., 3.

44. Ibid., 4.

45. Ibid., 6.

46. Francine Masiello, *Lenguaje e ideología: Las escuelas argentinas de vanguardia* (Buenos Aires: Hachette, 1986), 39.

47. Lindstrom, *Jewish Issues in Argentine Literature*, 7.

48. Dalia Kanidyoti, "Comparative Diasporas: The Local and the Mobile in Abraham Cahan and Alberto Gerchunoff," *Modern Fiction Studies* 44 (1998): 77–122.

49. Sarlo and Altamirano, "La Argentina del centenario," 74.

50. Nicholas Shumway, *The Invention of Argentina* (Berkeley: University of California Press, 1991), 29.

51. *Jewish Gauchos*, 42.

52. Lindstrom, *Jewish Issues in Argentine Literature*, 31.

53. See Jeffrey Lesser, *Negotiating National Identity: Immigrants, Minorities, and the Struggle for Ethnicity in Brazil* (Durham, NC: Duke University Press, 1999).

54. Sergio Pujol, *Las canciones del inmigrante: Buenos Aires, espectáculo musical y proceso inmigratorio de 1914 a nuestros días* (Buenos Aires: Almagesto, 1989), 41.

55. Lindstrom, *Jewish Issues in Argentine Literature*, 51.

56. Svetlana Boym explains that nostalgia is in itself a historically framed emotion. Thus, nostalgia is not diametrically opposed to modernity and individual responsibility; rather, it is a result of modernity itself. See her *The Future of Nostalgia* (New York: Basic Books, 2001), especially xvi.

57. Doris Sommer suggests that the function of *costumbrismo* was "to promote communal imaginings through the middle stratum of writers and
readers who constituted the most authentic expression of national feeling." See her *Foundational Fictions: The National Romances of Latin America* (Berkeley: University of California Press, 1991), 14. A characteristic
strategy to add local flavor was the use of regional words, something that
Gerchunoff took pains to avoid. In *Los gauchos judíos*, Yiddish words and
expressions are remarkably absent.

Urbanization and mass emigration brought together all sorts of languages, dialects, and idiolects previously separated by space and social difference. A defense of the language, such as that of the "eloquent" writers
criticized by Alcântara Machado, also becomes a way of defending the borders, those outlying borders crossed by foreigners and a growing urban working class. On the other hand, an insistence on dialect could also portend a
fracturing or radical and terrifying mutation of the Portuguese language and
of society as a whole. This Italian-Portuguese seems to be everywhere, in the
trolleys and in *Fanfulla*. Immigrants do not adapt to a supposed Brazilian
national identity, they transform it much as they are transformed by it. And
just as industry must progress, the *Modernista* writer had better embrace this
new world or become a fossil of the previous century.

CHAPTER 4

1. For a more detailed discussion, see Nancy Stepan, *The Hour of Eugenics:
Race, Gender, and Nation in Latin America* (Ithaca, NY: Cornell University Press, 1991), 88–89.
2. Jeffrey Lesser, "The Hidden Hyphen," in *Negotiating National Identity:
Immigrants, Minorities, and the Struggle for Ethnicity in Brazil* (Durham,
NC: Duke University Press, 1999), 1–12.
3. Sílvio Romero, *O allemanismo no sul do Brasil, seus perigos e méios de os
conjurar* (Rio de Janeiro: Heitor Ribeiro, 1906). For more on Romero,
see Antônio Cândido, *Sílvio Romero: teoria, crítica e história literária*
(Rio de Janeiro: Livros Técnicos e Científicos Editora, 1978).
4. Giralda Seyferth, "Imigração no Brasil: Os preceitos de exclusão,"
http://www.comciencia.br/reportagens/migracoes/migr03.htm.
5. Sílvio Romero, *O allemanismo no sul do Brasil, seus perigos e méios de os
conjurar* (Rio de Janeiro: Heitor Ribeiro, 1906), 1–3.
6. Ibid., 7.
7. Ibid., 8.
8. Ibid., 9 (italics in original).
9. Ibid., 10.
10. Ibid., 11.
11. Ibid., 35–36.

12. Ibid., 45 (italics in original).
13. Alberto Torres, *O problema nacional brasileiro. Introdução a um programa de organização nacional*, ed. Francisco Iglésias (São Paulo: Editora Nacional, 1982).
14. In his preface, Francisco Iglésias describes Torres's life in a more detailed fashion. See v–vii; the constitution incident is mentioned on v.
15. Torres, *O problema*, 14.
16. Ibid., 17.
17. Ibid., 29.
18. Ibid., 53.
19. Ibid., 64.
20. Ibid., 68–70.
21. Ibid., 69.
22. Bio-bibliographical material on Graça Aranha abounds, yet it is mostly descriptive in nature. One of the few articles in English that recognizes the importance of racial thought in the novel is Marshall Eakin, "Race and Ideology in Graça Aranha's *Canaã*," *Ideologies and Literature: Journal of Hispanic and Lusophone Discourse Analysis* 3, no. 14 (1980): 3–15.
23. José Pereira da Graça Aranha, *Canãa* (Rio de Janeiro: Instituto Nacional do Livro, 1969), 79.
24. Ibid., 70.
25. Ibid., 64.
26. Ibid., 88.
27. Ibid., 67.
28. Ibid., 196.
29. Ibid., 105.
30. Ibid., 194.
31. Ibid., 48–49.
32. Ibid., 49.
33. Ibid., 53.
34. Ibid., 136.
35. Dirce Côrtes Riedel, *Graça Aranha; introdução e notas biobibliográficas* (Rio de Janeiro: Ediouro, 1988).
36. Freyre himself suggests as much in his preface: "ainda hoje é assunto que . . . inquieta o Brasileiro: estará ou não assimilando . . . as culturas diversas que veêm entrando, com maior impacto, na sua composição?" (Still today it is the topic that . . . bothers the Brazilian: will he assimilate or not . . . the diverse cultures that continue entering, with greater impact, on his composition?) Gilberto Freyre, *Casa-grande & senzala* (Rio de Janeiro: José Olympio, 1958), 17.
37. Lesser, *Negotiating National Identity*, 8.
38. Statistics cited in Julio Ramos, "Faceless Tongues: Language and Citizenship in Nineteenth-Century Latin America," in *Displacements*:

Cultural Identities in Question, ed. Angelika Bammer (Bloomington: Indiana University Press, 1994), 25–46.

39. Zuleika Alvim, "Imigrantes: A vida privada dos pobres no campo," in *História da vida privada no Brasil*, vol. 2, *Império: A corte e a modernidade nacional*, ed. Ferando A. Novais (São Paulo: Companhia das Letras, 1997), 216–87.

40. Edward Timms and David Kelly, eds., "Introduction: Unreal City-Theme and Variations," in *Unreal City: Urban Experience in Modern European Literature and Art* (New York: St. Martin's Press, 1985), 1–12.

41. Nicolau Sevcenko, *Orfeu extático na metrópole: São Paulo, sociedade e cultura nos frementes anos 20* (São Paulo: Companhia das Letras, 1992), 37.

42. Statistics in David Aliano, "Brazil through Italian Eyes: The Debate over Emigration to São Paulo during the 1920s, "*AltreItalie* 2(2005): 20, 87–107.

43. Ibid., 105.

44. Benito Mussolini, quoted by Aliano, "Brazil Through Italian Eyes," 89.

45. Thomas Holloway, *Immigrants on the Land: Coffee and Society in São Paulo, 1886–1934* (Chapel Hill: University of North Carolina Press, 1980), 168.

46. Sevcenko, *Orfeu*, 37–38.

47. Joseph Love, *São Paulo in the Brazilian Federation 1889–1937* (Stanford, CA: Stanford University Press, 1980).

48. Ibid., 331.

49. Apparently, radical or foreign language newspapers were occasionally attacked. *Fanfulla* barely escaped the same fate, according to Maurício Font, *Coffee, Contention, and Change in the Making of Modern Brazil*, (Cambridge, MA: Basil Blackwell, 1990), 212.

50. Sevcenko, *Orfeu*, 37–38.

51. For a complete history of the Liga nacionalista, see Sílvia Moreira, *São Paulo na Primeira República* (São Paulo: Brasiliense, 1988), 137–139.

52. Lúcia Lippi, *O Brasil dos imigrantes* (Rio de Janeiro: Jorge Zahar, 2001).

53. Susan Besse, *Restructuring Patriarchy: The Modernization of Gender Inequality in Brazil 1914–1940* (Chapel Hill: University of North Carolina Press, 1996), 15.

54. Ibid., 22.

55. Quoted in Besse, *Restructuring Patriarchy*, 37.

56. Sevcenko, *Orfeo*, 137.

57. The spelling of the neighborhood varies. In order to make it clear that it is indeed the same neighborhood that is mentioned by Machado and *O estado de São Paulo*, I have homogenized its use to Bráz, the way it is currently spelled in São Paulo.

58. Quoted by Sevcenko, *Orfeu*, 129.

59. Alcântara Machado, quoted in Sevcenco, *Orfeu*, 118.

60. Ibid.

61. Ibid.

62. Eliezer Pacheco, *O partido comunista brasileiro* (São Paulo: Alfa-Omega, 1984), 40–45.

63. Sevcenko, *Orfeo*, 143–44.

64. *Em memória de Antônio de Alcântara Machado*, ed. Agripino Grieco (São Paulo: Alvino Pocai, 1936), 182.

65. See Michael North, "Against the Standard, Linguistic Innovation, Racial Masquerade, and the Modernist Rebellion," in *The Dialect of Modernism, Race, Language, and Twentieth-Century Literature* (New York: Oxford University Press, 1994), 3–36.

66. Antônio de Alcântara Machado, *Prosa preparatória & cavaquinho e saxophone*, vol. 1, ed. Francisco de Assis and Cecília de Lara (Rio de Janeiro: Civilização brasileira, 1983), 162. This volume contains letters as well as more theoretical essays on *Modernismo* aesthetics.

67. Mário Carelli, *Carcamanos e comendadores*, 192.

68. Antônio de Alcântara Machado, *Novelas paulistanas: Bráz, Bexiga e Barra Funda* (Rio de Janeiro: Ediouro, 2004), 9.

69. Antônio de Alcântara Machado, "Carmela," in *Novelas paulistanas: Bráz, Bexiga e Barra Funda*, 14.

70. Antônio de Alcântara Machado, "O monstro de rodas," in *Novelas paulistanas: Bráz, Bexiga e Barra Funda*, 47.

71. Heloísa Buarque de Hollanda, "Os estudos sobre mulher e literatura no Brasil: Uma primeira Avaliação," in *Uma questão de gênero*, ed. Albertina de Oliveira Costa and Cristina Bruschini (São Paulo: Fundação Carlos Chagas, 1992), 54–92.

72. Ibid., 88.

73. Vera Maria Chalmers, "Virado à paulista," in *Os pobres na literatura brasileira*, ed. Roberto Schwarz (Rio de Janeiro: Brasiliense, 1983) 136–39.

74. Antônio de Alcântara Machado, "Artigo de Fondo," in *Novelas paulistanas: Bráz, Bexiga e Barra Funda*, 9.

75. Ibid.

Bibliography

Primary Sources

Almeida, Júlia Lopes, and Francisco Felinto de Almeida [A. Julinto, pseud.]. *A casa verde*. São Paulo: Companhia Nacional, 1932.

Aranha, José Pereira da Graça. *Canãa*. Rio de Janeiro: Instituto Nacional do Livro, 1969.

Azevedo, Aluísio de. *O cortiço*. http://www.cervantesvirtual.com/servlet/SirveObras/05815063290570695209079/index.htm (accessed December 10, 2009). Translated by David H. Rosenthal as *The Slum* (London: Oxford University Press, 2001).

de la Barra de Llanos, Emma [César Duayen, pseud.]. *Stella (vida de costumbres argentinas)*. Barcelona, Editorial Maucci, 1908.

Gerchunoff, Alberto. *Los gauchos judíos*. Buenos Aires: Aguilar, 1975. Translated by Prudencio de Pereda as *The Jewish Gauchos of the Pampas* (Albuquerque: University of New Mexico Press, 1998).

Machado, Antônio de Alcântara. *Novelas paulistanas: Brás, Bexiga, e Barra Funda*. Rio de Janeiro: José Olympio, 1971.

Miró, José María [Julián Martel, pseud.]. *La bolsa*. Buenos Aires: Imprima editores, 1979.

Prado, Eduardo. *A ilusão americana*. São Paulo: Editora Brasiliense, 1961.

Rojas, Ricardo. *La restauración nacionalista*. Buenos Aires: Ministerio de justicia e instrucción pública, 1922.

Romero, Sílvio. *O allemanismo no sul do Brasil, seus perigos e méios de os conjurar*. Rio de Janeiro: Heitor Ribeiro, 1906.

Torres, Alberto. *O problema nacional brasileiro: Introdução a um programa de organização nacional*. 4th ed. São Paulo: Editora Nacional, 1982.

Souza, João Cardoso de Menezes. *Theses sobre colonização do Brazil: Projecto de solução as questões sociais, que se prendem à este difícil problema*. Rio de Janeiro: Typografia nacional, 1875.

De Zettery, Arrigo. *Manual del emigrante italiano*. Translated and edited by Diego Armus. Buenos Aires: Centro Editor de América Latina, 1983.

SECONDARY SOURCES

Acree, William. "Tracing the Ideological Line: Philosophies of the Argentine Nation from Sarmiento to Martínez Estrada." *A contracorriente* 1, no. 1 (2003): 102–33.

Ades, Dawn. "Modernism and the Search for Roots." In *Art in Latin America: The Modern Era, 1820–1980,* edited by Dawn Ades. New Haven, CT: Yale University Press, 1989.

Alencastro, Luiz Felipe, and Maria Luiza Renaux. "Caras e modos dos migrantes e imigrantes." In *História da vida privada no Brasil.* Vol. 2 of *Império: A corte e a modernidade* nacional, edited by Fernando Novais, 292–335. São Paulo: Compania de Letras, 1997.

Altamirano, Carlos and Beatriz Sarlo. "La fundación de la literatura argentina." In *Ensayos argentinos: De Sarmiento a la vanguardia.* Buenos Aires: Centro Editor de America Latina, 1983, 201–10.

Alvim, Zuleika. "Imigrantes: Avida privada dos pobres no campo." In *História da vida privada no Brasil.* Vol. 3 of *República: Da belle époque à era do rádio,* edited by Fernando A. Novais, 216–335. São Paulo: Companhia das Letras, 1997.

Aliano, David. "Brazil through Italian Eyes: The Debate over Emigration to São Paulo during the 1920s." *Altreitalie* 2 (2005): 87–107.

Anderson, Benedict. *Imagined Communities.* London: Verso, 1991.

Armstrong, Nancy. *Desire and Domestic Fiction: A Political History of the Novel.* New York: Oxford University Press, 1987.

Baily, Samuel. "The Role of Two Newspapers in the Assimilation of Italians in Buenos Aires and São Paulo, 1893–1913." *The International Migration Review* 12, no. 3 (1978): 321–40.

———. "Las sociedades de ayuda mutua y el desarrollo de una comunidad italiana en Buenos Aires 1858–1918." *Desarrollo económico* 21, no. 84 (1982): 485–514.

Balibar, Etiénne, and Immanuel Wallerstein. *Race, Nation, Class.* London: Verso, 1991.

Barrancos, Dora. *Anarquismo, educación y costumbres en la Argentina de principios de siglo.* Buenos Aires: Editorial Contrapunto, 1990.

Berg, Mary. Introduction to *Stella: Una novela de costumbres argentinas,* by César Duáyen. Buenos Aires: Stockcero, 2005.

Berman, Marshall. "Modernity—Yesterday, Today and Tomorrow." In *All That is Solid Melts into Air: The Experience of Modernity,* 15–36. New York: Penguin, 1988.

Bertoni, Lilia Ana. "La naturalización de los extranjeros 1887–1893: ¿Derechos políticos o nacionalidad?" *Desarrollo Económico* 32, no. 125 (1992): 57–77.

Besse, Susan. *Restructuring Patriarchy: The Modernization of Gender Inequality in Brazil, 1914–1940*. Chapel Hill: University of North Carolina Press, 1996.

Bletz, May E. "Agrarian Utopia in Alberto Gerchunoff's Los gauchos judíos." *Brújula* 3, no. 1 (2004): 35–52.

———. "Júlia Lopes de Almeida." In *Latin American Women Writers: An Encyclopedia*, edited by Eva Bueno and María Claudia André, 23–25. New York: Routledge, 2007.

———. "Race and modernity in *O cortiço* by Aluísio de Azevedo" *LLJournal* 2, no. 1 (2007). http://ojs.gc.cuny.edu/index.php/lljournal/article/view/231/198 (accessed December 10, 2009).

Le Bon, Gustave. *Psychologie des foules*. 4th ed. Paris: Presses universitaires de France, 1991.

Borges, Dain. "'Puffy, Ugly, Slothful, and Inert': Degeneration in Brazilian Social Thought, 1880–1940." *Journal of Latin American Studies* 25 (1993): 235–56.

Boym, Svetlana. *The Future of Nostalgia*. New York: Basic Books, 2001.

Burchell, Graham, Colin Gordon, and Peter Miller, eds. *The Foucault Effect: Studies in Governmentality, with two Lectures by and an Interview with Michel Foucault*. Chicago: University of Chicago Press, 1991.

Cândido, Antônio. "El paso del dos al tres (contribución al estudio de las mediaciones en el análisis literario)" *Escritura: Teoría y Critica Literarias* 3 (1977): 21–34.

———. *Sílvio Romero: Teoria, crítica e história literária*. Rio de Janeiro: Livros Técnicos e Científicos Editora, 1978.

Carelli, Mário. *Carcamanos e comendadores*. São Paulo: Editora Ática, 1985.

Chalmers, Vera Maria. "Virado à paulista." In *Os pobres na literatura brasileira*, edited by Roberto Schwarz, 136–39. Rio de Janeiro: Brasiliense, 1983.

Chasteen, John Charles, and Sara Castro-Klarén, eds. *Beyond Imagined Communities: Reading and Writing the Nation in Nineteenth-Century Latin America*. Baltimore, MD: Johns Hopkins University Press, 2004.

Chazkel, Amy. "The Crônica, the City, and the Invention of the Underworld: Rio de Janeiro, 1889–1922." *Estudios interdisciplinarios de América Latina y el Caribe* 12, no. 1 (2001): 79–105.

Civantos, Cristine Elsa. *Between Argentines and Arabs: The Writing of National and Immigrant Identities*. Albany: State University of New York Press, 2006.

Clifford, James. "Diasporas." *Cultural Anthropology* 9, no. 3 (1994): 302–38.

Delaney, Jeane. "Making Sense of Modernity: Changing Attitudes toward the Immigrant and the Gaucho in Turn-of-the-Century Argentina." *Comparative Studies in Society and History* 38, no. 3 (1996): 434–59.

―――. "National Identity, Nationhood, and Immigration in Argentina: 1810–1930." *Stanford Electronic Humanities Review* 5, no. 2 (1997), http://stanford.edu/group/SHR/5-2/delaney.html (accessed December 10, 2009).

Devoto, Fernando. *Movimientos migratorios: Historiografía y problemas.* Buenos Aires: Centro Editor de América Latina, 1992.

Eakin, Marshall. "Race and Ideology in Graça Aranha's *Canaã.*" *Ideologies and Literature: Journal of Hispanic and Lusophone Discourse Analysis* 3, no. 14 (1980): 3–15.

Fausto, Boris. *Fazer a América: A imigração em massa para a América Latina.* São Paulo: Edusp, 1999.

Felski, Rita. *The Gender of Modernity.* Cambridge, MA: Harvard University Press, 1995.

Font, Maurício. *Coffee, Contention, and Change in the Making of Modern Brazil.* Cambridge, MA: Basil Blackwell, 1990.

Foucault, Michel. *The History of Sexuality: An Introduction.* Vol. 1. New York: Vintage, 1988.

―――. *"Society Must Be Defended": Lectures at the Collège de France, 1975–76.* London: Penguin Books, 2003.

Frederick, Bonnie. *Wily Modesty: Argentine Women Writers, 1860–1910.* Tempe: Arizona State University Press, 1998.

Freyre, Gilberto. *Casa-grande & senzala.* 9th ed. Rio de Janeiro: José Olympio, 1958.

―――. *Sobrados e mucambos; decadência do patriarcado rural e desenvolvimento do urbano.* 4th ed. Rio de Janeiro, José Olympio, 1968.

Glaubert, Earl. "Ricardo Rojas and the Emergence of Argentine Cultural Nationalism." *Hispanic American Historical Review* 43, no. 1 (1963): 1–13.

Guenther, Louise. "The British Community of 19th Century Bahia: Public and Private Lives." Working paper, University of Oxford Centre for Brazilian Studies Working Paper Series. http://www.brazil.ox.ac.uk/__data/assets/pdf_file/0003/9399/Guenther32.pdf (accessed December 10, 2009).

―――. *British Merchants in Nineteenth-Century Brazil: Business, Culture, and Identity in Bahia, 1808–1850.* Oxford: Oxford University Press, 2004.

Halperín Donghi, Tulio. "Argentine Counterpoint: Rise of the Nation, Rise of the State." In *Beyond Imagined Communities: Reading and Writing the Nation in Nineteenth-Century Latin America,* edited by Sara Castro-Klarén and John Charles, 33–53. Baltimore, MD: Johns Hopkins University Press, 2003.

———. "¿Para qué la inmigración? Ideología y política inmigratoria y aceleración del proceso modernizador: El caso argentine (1810–1914)." In *El espejo de la historia. Problemas argentinos y perspectivas Latinoamericanas*, 189–238. Buenos Aires: Sudamericana, 1987.

Hollanda, Heloísa Buarque. "Os estudos sobre mulher e literatura no Brasil: Uma primeira avaliação." In *Uma questão de gênero*, edited by Albertina de Oliveira Costa and Cristina Bruschini, 54–92. São Paulo: Fundação Carlos Chagas, 1992.

Holloway, Thomas. *Policing Rio de Janeiro: Representation and Resistance in a 19th-Century City*. Stanford, CA: Stanford University Press, 1993.

———. *Immigrants on the Land: Coffee and Society in São Paulo, 1886–1934*. Chapel Hill: University of North Carolina Press, 1980.

Honig, Bonnie. "Ruth, the Model Émigré: Mourning and the Symbolic Politics of Immigration." In *Cosmopolitics: Thinking and Feeling Beyond the Nation*, edited by Cheah Pheng and Bruce Robbins,192–215. Minneapolis: Minneapolis University Press, 1998.

Jacobson, Matthew Fry. *Whiteness of a Different Color: European Immigrants and the Alchemy of Race*. Cambridge, MA: Harvard University Press, 1998.

Kandiyoti, Dalia. "Comparative Diasporas: The Local and the Mobile in Abraham Cahan and Alberto Gerchunoff." *Modern Fiction Studies* 44 (1998): 77–122.

Kaplan, Caren. *Questions of Travel: Postmodern Discourses of Displacement*. Durham, NC: Duke University Press, 1996.

Korn, Francis. "Algunos aspectos de la asimilación de inmigrantes en Buenos Aires." In *Los fragmentos del poder*, edited by Torcuato di Tella, 441–60. Buenos Aires: Jorge Alvarez, 1969.

Lesser, Jeffrey. *Negotiating National Identity: Immigrants, Minorities, and the Struggle for Ethnicity in Brazil*. Durham, NC: Duke University Press, 1999.

Lindstrom, Naomi. *Jewish Issues in Argentine Literature: From Gerchunoff to Szichman*. Columbia: University of Missouri Press, 1989.

Love, Joseph. *São Paulo in the Brazilian Federation 1889–1937*. Stanford, CA: Stanford University Press, 1980.

Lowe, Lisa. *Immigrant Acts: On Asian American Cultural Politics*. Durham, NC: Duke University Press, 1996.

Machado, Antônio de Alcântara. *Prosa preparatória & cavaquinho e saxofone*. Vol. 1, edited by Francisco de Assis Barbosa and Cecília de Lara. Rio de Janeiro: Civilização brasileira, 1983.

Marchant, Elizabeth. "Naturalism, Race, and Nationalism in Aluísio Azevedo's *O mulato*." *Hispania* 83, no. 3 (2000): 445–53.

Masiello, Francine. *Between Civilization and Barbarism: Women, Nation, and Literary Culture in Modern Argentina*. Lincoln: University of Nebraska Press, 1992.

———. *Lenguaje e ideología: Las escuelas argentinas de vanguardia*. Buenos Aires: Hachette, 1986.

Mérian, Jean-Yves. *Aluísio Azevedo: Vida e obra. O verdadeiro Brasil do século XIX*. Rio de Janeiro: Espaço e tempo, 1988.

McClintock, Anne. "Family Feuds: Gender, Nationalism, and the Family." *Feminist Review* 44 (1993): 61–80.

Mead, Karen. "Gendering the Obstacles to Progress in Positivist Argentina 1880–1920." *Hispanic American Historical Review* 77, no. 4 (1997): 645–75.

Mirelman, Víctor A. *Jewish Buenos Aires, 1890–1930: In Search of an Identity*. Detroit, MI: Wayne State University Press, 1990.

Molloy, Sylvia. "Madre patria y la madrastra: Figuración de España en la novela familiar de Sarmiento." *La Torre* 1, no. 1 (1987): 45–58.

Moreira, Sílvia. *São Paulo na Primeira República*. São Paulo: Brasiliense, 1988.

Moya, José. *Cousins and Strangers*. Berkeley: University of California Press, 1998.

Needell, Jeffrey. *A Tropical Belle Époque: Elite Culture and Society in Turn-of-the-Century Rio de Janeiro*. Cambridge: Cambridge University Press, 1987.

North, Michael. *The Dialect of Modernism: Race, Language, and Twentieth-Century Literature*. New York: Oxford University Press, 1994.

Nouzeilles, Gabriela. "Pathological Romances and National Dystopias in Argentine Naturalism." *Latin American Literary Review* 24, no. 47 (1996): 23–39.

Outtes, Joel. "Disciplining Society through the City: The Genesis of City Planning in Brazil and Argentina (1894–1945)." *Bulletin of Latin American Research* 22, no. 2 (2003): 137–64.

Oliveira, Lucia Lippi. *O Brasil dos imigrantes*. Rio de Janeiro: Jorge Zahar, 2001.

Onega, Gladys. *La inmigración en la literatura argentina (1880–1910)*. Buenos Aires: Centro Editor de América Latina, 1982.

Pacheco, Eliezer. *O partido comunista brasileiro*. São Paulo: Alfa-Omega, 1984.

Peard, Julyan. *Race, Place, and Medicine: The Idea of the Tropics in Nineteenth-Century Brazilian Medicine*. Durham, NC: Duke University Press, 1999.

Pereira, Lúcia Miguel. *História da literatura brasileira: Prosa da ficção*. Rio de Janeiro: José Olympia, 1973.

Pratt, Mary Louise. "Women, Literature, and the National Brotherhood." *Nineteenth-Century Contexts* 18, no. 1 (1994): 27–47.

Puiggrós, Adriana. *Sujetos, disciplina y curriculum: En los orígenes del sistema educativo argentino*. Buenos Aires: Editorial Galerna, 1990.

Pujol, Sergio. *Las canciones del inmigrante: Buenos Aires: Espectáculo musical y proceso inmigratiorio, de 1914 a nuestros días*. Buenos Aires: Almagesto, 1989.

Ramos, Julio. "Faceless Tongues: Language and Citizenship in Nineteenth-Century Latin America." In *Displacements: Cultural Identities in Question*, edited by Angelika Bammer, 25–46. Bloomington: Indiana University Press, 1994.

Ribeiro, Gladys Sabina. *Mata galegos: Os portugueses e os conflitos de trabalho na República Velha*. São Paulo: Brasiliense, 1989.

Riedel, Dirce Côrtes. *Graça Aranha; introdução e notas biobibliográficas*. Rio de Janiero: Ediouro, 1988.

Rio, João do. "Um lar de Artistas," in *Momento Literário* (Rio de Janeiro: Fundação Biblioteca Nacional, 1994), http://www.cervantesvirtual.com/servlet/SirveObras/12693858724598273098435/p0000001.htm#I_5 (accessed Dec. 10, 2009).

Rock, David. "From the First World War to 1930." In *Argentina Since Independence*, edited by Leslie Bethell, 139–72. Cambridge: Cambridge University Press, 1993.

———. "Intellectual Precursors of Conservative Argentine Nationalism." *Hispanic American Historical Review* 67, no. 2 (1987): 271–300.

Rodrigues, Raymundo Nina. *As raças humanas e a responsabilidade penal no Brasil*. São Paulo: Companhia Editora Nacional, 1938.

Romero, José Luis. *Las ideologías de la cultura nacional y otros ensayos*. México: Fondo de Cultura Económica, 1965.

Romero, Luis Alberto. "Buenos Aires en la entreguerra: Libros baratos y culturas de los sectores populares." In *Mundo urbano y cultura popular: Estudios de historia social argentina*, editcd by Diego Armus, 39–67. Buenos Aires: Sudamericana, 1990.

Roncador, Sônia. *A doméstica imaginária: Literatura, testemunhos, e a invenção da empregada doméstica no Brasil (1889–1999)*. Brazil: Editora Universidade de Brasília, 2008.

———. "O demônio familiar: Lavadeiras, amas-de-leite e criadas na narrativa de Júlia Lopes de Almeida." *Luso-Brazilian Review* 44, no. 1 (2007): 94–121.

Rosaldo, Renato. *Culture and Truth: The Remaking of Social Analysis*. Boston, MA: Beacon Press, 1989.

Sábato, Hilda. *The Many and the Few: Political Participation in Republican Buenos Aires*. Stanford, CA: Stanford University Press: 2001.

Salvatore, Ricardo, and Carlos Aguirre, eds. *The Birth of the Penitentiary in Latin America: Essays on Criminology, Prison Reform, and Social Control, 1830–1940*. Austin: University of Texas Press, 1996.

Said, Edward. *Culture and Imperialism*. New York: Columbia University Press, 1993.

———. *The World, the Text, and the Critic*. Cambridge, MA: Harvard University Press, 1983.

Sant'Anna, Afonso Romano de. "Curtição: O cortiço do mestre Cândido e o meu." *Minas Gerais, Suplemento Literario* 2 (April 16, 1977): 6–9.

Sarlo, Beatriz. "La Argentina del centenario." In *Ensayos argentinos: De Sarmiento a la vanguardia*, edited by Carlos Altamirano and Beatriz Sarlo, 69–105. Buenos Aires: Centro Editor de America Latina, 1983.

———. *Una modernidad periférica*. Buenos Aires: Nueva Visión, 1988.

———. "Modernidad y mezcla cultural. El caso de Buenos Aires." In *Modernidade: Vanguardas artísticas na América Latina*, edited by Ana Maria de Moraes Belluzzo, 31–43. São Paulo: Fundação Memorial da América Latina, 1990.

———. "Oralidad y lenguas extranjeras. El conflicto en la literatura argentina durante el primer tercio del siglo XX." In *Oralidad y argentinidad: Estudios sobre la functión del lenguaje hablado en la literatura argentina*, edited by Walter Bruno Berg and Markus Klaus Schaffauer, 28–41. Tübingen, Germany: Narr, 1997.

Schneider, Arnd. "The Transcontinental Construction of European Identities: A View from Argentina." *Anthropological Journal on European Cultures* 5, no. 1 (1996): 95–105.

Schwarz, Roberto. "Nacional por subtração." In *Que horas são*? São Paulo: Companhia das Letras, 1987, 29–48. Translated by Roberto Schwarz as "Brazilian Culture: Nationalism by Elimination." (*New Left Review* [1988]: 77–90).

Senkman, Leonardo. "*Los gauchos judíos*: Una lectura desde Israel." *Estudios interdisciplinarios de América Latina* 10, no. 1 (1999): 141–52.

Sevcenko, Nicolau. *Orfeu extático na metrópole: São Paulo, sociedade e cultura nos frementes anos 20*. São Paulo: Companhia das Letras, 1992.

Seyferth, Giralda. "Imigração no Brasil: Os preceitos de exclusão." http://www.comciencia.br/reportagens/migracoes/migr03.htm (accessed December 10, 2009).

Sharpe, Peggy. "O caminho crítico d'*A viúva Simões*." In *A Viúva Simões*, by Júlia Lopes de Almeida, 9–26. Florianópolis: Editora Mulheres, 1999.

Shumway, Nicolas. *The Invention of Argentina*. Berkeley: University of California Press, 1991.

Simmel, Georg. "Metropolis and Mental Life." In *On Individuality and Social Forms*, edited by Donald N. Levine and Morris Janowitz, 143–49. Chicago: University of Chicago Press, 1971.

———. "The Stranger." In *On Individuality and Social Forms*, edited by Donald N. Levine and Morris Janowitz, 324–39. Chicago: University of Chicago Press, 1971.

Skidmore, Thomas. *Black into White: Race and Nationality in Brazilian Thought*. Durham, NC: Duke University Press, 1995.

Solberg, Carl. *Immigration and Nationalism, Argentina and Chile 1890–1914*. Austin: University of Texas Press, 1970.

Soler, Ricaurte. *El positivismo argentino: Pensamiento filosófico y sociológico*. México: Universidad Nacional Autónoma de México, Facultad de Filosofía y Letras, 1979.

Sommer, Doris. *Foundational Fictions: The National Romances of Latin America*. Berkeley: University of California Press, 1991.

Sorensen, Diana. "Ricardo Rojas, lector del *Facundo*: Hacia la construcción de la cultura nacional." *Filología* 21, no. 1 (1986): 173–81.

Stepan, Nancy. *The Hour of Eugenics: Race, Gender, and Nation in Latin America*. Ithaca, NY: Cornell University Press, 1991.

Stoler, Ann Laura. *Race and the Education of Desire: Foucault's History of Sexuality and the Colonial Order of Things*. Durham, NC: Duke University Press, 1995.

Suriano, Juan. *La huelga de inquilinos de 1907*. Buenos Aires: Center for the Economic Analysis of Law, 1983.

Svampa, Marisella. "Inmigración y nacionalidad." *Studi Emigrazioni/Études Migrations* 30 (1993): 289–310.

Tedesco, Juan Carlos. *Educación y sociedad en la Argentina: 1880–1945*. Buenos Aires: Ediciones Solar, 1986.

Tella, Torcuato di. "Raíces de la controversia educacional argentina." In *Los fragmentos del poder*, edited by Torcuato di Tella, 289–323. Buenos Aires: Jorge Alvarez, 1969.

Terán, Oscar. *Positivismo y nación en la Argentina con una selección de textos de J. M. Ramos Mejía, Con una selección de textos de J. M. Ramos Mejía, A. Alvarez, C. O. Bunge y J. Ingenieros*. Buenos Aires: Punto Sur, 1987.

Timms, Edward, and David Kelly, eds. *Unreal City: Urban Experience in Modern European Literature and Art*. New York: St. Martin's Press, 1985.

Unruh, Vicky. *Latin American Vanguards*. Berkeley: University of California Press, 1994.

Vezzetti, Hugo. *La locura en la Argentina*. Buenos Aires: Folios, 1983.

Williams, Raymond. *The Country and the City*. Oxford: Oxford University Press, 1973.

————. "Metropolitan Perceptions and the Emergence of Modernism." In *The Politics of Modernism: Against the New Conformists*, by Raymond Williams, 37–47. London: Verso, 1989.

Zimmermann, Eduardo. "Racial Ideas and Social Reform: Argentina, 1890–1916." *Hispanic American Historical Review* 72, no. 1 (1992): 23–46.

INDEX